INTERDISCIPLINARY ESSAYS ON DARWINISM IN HISPANIC LITERATURE AND FILM

The Intersection of Science and the Humanities

INTERDISCIPLINARY ESSAYS ON DARWINISM IN HISPANIC LITERATURE AND FILM

The Intersection of Science and the Humanities

Edited by
Jerry Hoeg
and
Kevin S. Larsen

With a Foreword by
Joseph Henry Vogel

The Edwin Mellen Press
Lewiston•Queenston•Lampeter

Library of Congress Cataloging-in-Publication Data

Interdisciplinary essays on Darwinism in Hispanic literature and film : the intersection of science and the humanities / edited by Jerry Hoeg and Kevin S. Larsen ; with a foreword by Joseph Henry Vogel.

p. cm.

Includes bibliographical references.

ISBN-13: 978-0-7734-4747-9

ISBN-10: 0-7734-4747-4

1. Latin American literature--History and criticism. 2. Spanish literature--History and criticism. 3. Evolution (Biology) in literature. 4. Biology in motion pictures. 5. Science and the humanities--Latin America. 6. Darwin, Charles, 1809-1882--Influence. I. Hoeg, Jerry, 1951- II. Larsen, Kevin S.

PQ7081.I55 2009

860.9'36--dc22

2009015892

hors série.

A CIP catalog record for this book is available from the British Library.

Front cover art: Montage by Patrick Daugherty (images from public domain)

The Edwin Mellen Press
Box 450
Lewiston, New York
USA 14092-0450

The Edwin Mellen Press
Box 67
Queenston, Ontario
CANADA L0S 1L0

The Edwin Mellen Press, Ltd.
Lampeter, Ceredigion, Wales
UNITED KINGDOM SA48 8LT

Printed in the United States of America

A Charles-Robert, in memoriam

Table of Contents

Foreword: Unintended Audiences

Who is the intended audience? The question is a primary one and the answer, non-obvious. From the title of this anthology, I am tempted to say "interdisciplinary academics in *both* the sciences and the humanities" but I do not yield to temptation. I can imagine the publisher's reaction: a sigh of exasperation. The market for interdisciplinary works is thin and its very thinness reflects the gap between the "two cultures" that C.P. Snow famously lectured about in 1959. Fifty years onward and the metaphor is still apt: many professors in the humanities/sciences do not even know where the departments of sciences/humanities are located on their own campus! Yet the reader of this anthology will discover that the sciences are indeed extending into the humanities and, *vice versa*, the humanities, into the sciences. The seeming paradox of more *and* less interdisciplinary work is resolved when we recognize the scale of academia today. The professorate is now so large that the number of eclectic scholars has risen even though their percentage has probably declined. Diffused across institutions, the hybrids collaborate through international symposia and the internet. Some boast unlikely combinations of academic degrees and many simultaneously publish in journals in each of the two cultures. Their scholarship ceases to be interdisciplinary and becomes transdisciplinary whenever it helps to re-configure the disciplines from which they have drawn their insights. An indicator of success is when a fellow academic remarks "I can't believe that he/she is in the Department of (fill in the blank)! I could have sworn that he/she was a (fill in the blank)." Such surprises await the reader of this anthology. So, to answer to my own question about the intended audience, I now respond: anyone who relishes the possibility of a Gestalt. The publisher can now breathe a sigh of relief. The market is immense.

In order to grapple with the sciences and the humanities, the editors Jerry Hoeg and Kevin S. Larsen have had to choose among the sciences and humanities. The science of choice is evolution and the humanities of choice, narratives. I am careful to say "evolution" and not "Darwinism" as the latter does not signal the whole body of scientific thought that has arisen since the debut of Darwin's foundational "descent with modification by means of natural selection" (exactly one hundred years before Snow's lecture). I am also careful to say "narratives" and not "literature" as that the latter would exclude the pre-literate oral traditions and contemporary filmic representations also analyzed in these chapters. By applying the science of evolution, and particularly of evolutionary psychology, to the analysis of narratives, the contributors are making sense of the humanities as well as suggesting new research streams. The fact that many of the themes explore idiosyncrasies, hitherto unexplained, coheres with Thomas Kuhn's structure of scientific revolutions. For example, Camilo Gomides, explains in evolutionary terms why Cervantes would have elevated his military status in an otherwise autobiographical story of his ordeal as a captive; David Barash, why predictable patterns emerge in "magical realism" despite the ostensible fantasy of the genre; and Joseph Carroll, why the re-examinations by Gomides, Barash and all the other contributors are "putting the human in its place."

Evolution is about "why?" and I confess to being obsessed with that question. My obsession is all the more peculiar inasmuch as I am an economist and my sub-culture instructs me *not* to ask "why?" with respect to large swaths of human behavior (e.g., consumer tastes, risk-taking, the work-*vs.*-leisure choice). We economists invoke the mantra of "sovereignty of preferences" and defer the "why" question to our brethren in the social sciences, viz., psychologists, sociologists, and anthropologists. Make no mistake. Our deference is not out of respect for the explanatory power of those disciplines or even out of an efficient division of mental labor. It is self-serving. By ignoring "why" we (1) suppress the meaninglessness of "equilibrium" when "demand" is manufactured by "supply" and (2) celebrate the market (i.e., demand and supply) as a beneficial mechanism to assign scarce resources. John Kenneth Galbraith exposed this sleight of hand in *The Affluent Society* about the same time C.P. Snow gave his "two cultures" lecture. Although Galbraith's book was a bestseller, the profession nevertheless ignored him and developed an edifice which teeters on *not* asking why. By *not* asking why, we now share an uneasy company with literary theorists whose establishment identifies itself with postmodernism. I say "uneasy company" because postmodernists accept cultural relativism as the ultimate truth and have, as E.O. Wilson puts it, "chosen not to encumber themselves with a scientific education" (44). Such elegant condescension cannot be applied to us the economists. We are trained to formulate refutable hypotheses and meticulously collect and analyze data. I predict that once we embrace the "why" question, the gates will fling open for evolutionary reasoning. My sense is that the door is already ajar as evident in the wildly popular *Freakanomics*. I also predict that through adaptive explanations of economic behavior, evolution may finally shake the humanities. An example can be found in this anthology. Michelle Scalise Sugiyama explains trickster stories as cognitive models of the free rider problem--a quintessentially economic behavior. After reading her chapter "I can't believe she is in the Department of Anthropology! I could have sworn she was an evolutionary psychologist" [in fact, she holds a Ph.D. in English].

Where does reduction end? If one can make sense of the humanities in the light of evolution, can one then make sense of evolution in the light of thermodynamics? And by transitivity, can one also make sense of the humanities in terms of thermodynamics? E.O. Wilson cites Niels Bohr that "The opposite of a correct statement is a false statement. But the opposite of a profound truth may be another profound truth" (1977, 127). Non-equilibrium thermodynamics appears to be the anti-discipline of evolution and both are profound truths. The pioneer of that revolutionary stance is Ilya Prigogine who won the 1977 Nobel Laureate in Chemistry. His *From Being to Becoming* explains the apparent contradiction of evolution as entropy by showing that increasing complexity of a system flows from its increasing the disorder of the surroundings. I had the good fortune to hear Prof. Prigogine lecture several years after I had worked up a mathematical model of evolution as an entropy-driven process (Vogel 1988). With an accent reminiscent of Inspector Poirot in an Agatha Christie novel-to-movie adaptation, Prigogine said that Newtonian physics is still used to calculate how to get a rocket to the moon--relativity theory would be overkill--we choose the science for the task at hand. Perhaps evolutionary psychology is all that is needed to make sense of the humanities. But my obsession with the why question inclines me to believe that other idiosyncrasies will be revealed when we reduce narratives all the way down to the anti-discipline.

I think about *Babel,* the award-winning film set in Morocco, California, Mexico, and Japan. In this anthology, David Michelson has brilliantly elucidated the empathy and conflicting moral norms evoked by that movie among self-identifying "liberal" or "conservative" viewers. Not surprising for an evolutionary psychologist, in-group allegiance and out-group dehumanization manifest themselves in responses to the dramatic events depicted. The dichotomy has an obvious adaptive value in our hominid past. From the viewpoint of non-equilibrium thermodynamics, its neurology is a structure that once dissipated the energy and material flows of a hunter, gatherer, and scavenger society. The prior sentence is not just a translation of one scientific language (evolution) into another (non-equilibrium thermodynamics). It does more. It invites further elucidation. The trilogy of non-equilibrium thermodynamics are fluctuations, boundary conditions, and amplification effects. In *Babel*, an adolescent goat herder shoots at a passing tour bus from a rugged hilltop in a desolate landscape. Unintentionally (fluctuation), the bullet penetrates an American passenger (boundary condition), which sets off a series of unfortunate events for the families of both the American and the Moroccan (boundary conditions). Such full reduction in the light of non-equilibrium thermodynamics will let us make even *better* sense of the narrative than just seeing it in the light of evolution.

Unintended audiences will greatly profit from this anthology.

Works Cited

Galbraith, John Kenneth. *The Affluent Society*. Boston: Houghton Mifflin, 1958.

Kuhn, Thomas. *The Structure of Scientific Revolutions*. Chicago: The U of Chicago P, 1962.

Levitt, Steven and Stephen J. Dubner. *Freakonomics: a Rogue Econonomist Explores the Hidden Side of Everything*. New York: William Morrow, 2005.

Prigogine, Ilya. *From Being to Becoming*. San Francisco: W.H. Freeman, 1980.

Snow, C.P. *The Two Cultures and the Scientific Revolution*. New York: Cambridge UP, 1959.

Vogel, Joseph Henry. "Evolution as an Entropy Driven Process: An Economic Model." *Systems Research* 5 (1988): 299-312.

Wilson, E.O. "The Concepts of Disciplines and Antidisciplines." *Daedalus*. September (1977): 127-140.

---. *Consilience: The Unity of Knowledge*. New York: Alfred A. Knopf, 1998

ACKNOWLEDGEMENTS

The editors would first like to thank the Ometeca Institute, where the idea for this study was born. We would also like to thank Eva Paulino Bueno, of St. Mary's University, San Antonio, for her excellent translation from the Portuguese of Chapter 8. We are also grateful to Joseph Vogel, of the University of Puerto Rico-Río Piedras, for writing the foreword to this study, and to Joseph Carroll of the University of Missouri-St. Louis, and Travis Landry of Kenyon College, for their invaluable aid in organizing the bibliography for this volume. And finally, thanks to the Pennsylvania State University for granting sabbatical time to Jerry Hoeg in order that he might have the time to complete this task.

DECLARATION

An earlier version of Chapter 6 was published in Spanish in *Ometeca* 9.1 (2005). An earlier version of Chapter 8 was published in Portuguese in *Ometeca* 12.1 (2008), as was an earlier version of Chapter 3.

INTRODUCTION

The occasions of this present volume are multiple. With regard to orders of magnitude, we first note in 2009 the second centenary of Charles Darwin's birth, as well as the sesquicentennial of the publication of the *Origin of Species*. Once his carefully-husbanded theory was loosed to the world, the English naturalist became almost immediately, and remains today, a larger-than-life public figure. He is undeniably one of the most influential—whether for good or for ill—figures in the intellectual history of our planet. His writings continue to figure among the most seminal tomes of science and, more recently, the study of the humanities. One of the salient characteristics of Darwinian thought is its versatility, even its malleability. From the publication of *The Origin* and *The Descent of Man,* Darwin's ideas (or what people thought were his ideas) have been interpreted and reinterpreted into a plethora of venues via a variety of vehicles. A number of the concepts propounded in his name, some gaining international currency, would surely have given Darwin pause, if not startled and even horrified him. Certain facets, even major ones, of Darwinism, whether in the later nineteenth century or at the beginning of the twenty-first, may bear scant resemblance to the original dimensions and doctrines of Darwin's work. From Huxley and Spencer, through Watson and Crick, down to Stephen Jay Gould and Niles Eldredge--these latter two the originators of the innately hybrid notion of punctuated equilibria--his followers have individually and collectively fostered a multifarious evolution within the movement itself. At the behest of creationists and strict constructionists alike, much of this development has occurred in an inescapably teleological manner, modifying the original tenets and even doctrines of the concept of "descent with modification" such that Darwin himself might scarcely recognize them as his disciplinary offspring.

Among the humanistic offshoots of the Darwinian model are studies of literature and film from an evolutionary or adaptationist perspective. Once again, such works vary according to circumstance, the author/critic, the subject matter, and the perspective and background of the reader. The engineering, not to mention the actual construction, of the bridges between the "two cultures" that C.P. Snow described, vary widely--even wildly--but the gap continues to be spanned, if not actually closed. While on this trajectory (tangent?) of unity in diversity, we also note another occasion worth commemorating: Ometeca (a word from the Nahuatl meaning "two in one"), an entity with which both editors have long been associated, marked its majority this past year. Twenty-one years ago last fall, Ometeca, first the journal and then the Institute, were conceived and then born (or hatched or whatever), beginning around a banquet table at the first meeting of the Society for Literature and Science in 1987, to focus on the interface of the sciences and the humanities in the Hispanic tradition. Under the able leadership of Dr. Rafael Catalá, the founding editor, manuscripts were gathered and edited; the first issue of the journal appeared in 1989. More have followed, as have meetings of the Institute in an array of countries in North and South America.

Along the way, the concepts we so enthusiastically and, on more than one occasion, so naively, espoused and promulgated to any and all who would listen (and to some who would not) have continued to evolve, sometimes scarcely bearing more than the grossest resemblance to the original line of thought and practice. What we thought in 1989 might be "definitive" became--thankfully!--formative and even nurturing, but was far from the "last word." In like measure, this present volume will certainly not be--not now nor ever--the very "last word" in Darwinian studies, not even in Darwinian literary studies, whether in the Hispanic world or anywhere else. Honestly, it's not even the "first word," nor was Ometeca. Hopefully, that ultimate statement will never be uttered, nor will the last word ever be spoken. In the humanities, as in the natural world, stasis, in literature as in life, and especially in scholarship, effectively equates to death. All living entities define themselves and have their being with regard to change and diversification. Any "definitive" statement can only reference dead (or dying) languages, dead (or dying) cultures, and dead (or dying) organisms.

Our humble assembly of essays ranges over relatively substantial expanses of time and space, over generations and genres of works and writers. We'd like to think that they offer a--not the--valid, if not an original, contribution to the ongoing discussion in which we are participants. We believe that the essays in this volume are solid scholarship, a trait that over time will almost always be selected for and, like Darwinism itself, should continue to be viable. Our hope is to make lasting contributions, fomenting, rather than truncating, evolution within and across the disciplines. We want only to open and to amplify dialogue, rather than to close it off. Likewise, literary and film studies have often fallen prey to the latest critical fads, purported paradigm shifts in theory that may make a loud noise for a time. But once their initial impetus is expended, they are consigned to the files of the more or less extinct or extinguished. We have attempted to let the texts and topics under discussion shine through, without a heavy theoretical overlay. It is our belief that scholarship should illuminate, rather than obfuscate. Any new approaches elaborated herein hopefully should not obscure the primary materials. Not that the critical can(n)on shouldn't be fired on occasion, if for no other reason than to stir things up. If we roil the waters just a bit, if we lob a round or two (or twelve) into the genetic pool that the humanities and the sciences share, if only along the margins, with a little luck we'll not immediately be relegated to the intellectual backwaters as froth, though time and tides will tell.

The Editors

CHAPTER 1

Magical Realism, Biological Reality, and the Hidden Biology of Fantasy, Secrecy, and Mystery

First, a confession: I am an evolutionary biologist, not an expert on the literature written in the Spanish language. Second, a bold (or perhaps, foolish) suggestion: perhaps the mainstay of the former discipline--evolution--sheds some light on one of the mainstays of the latter: magical realism. Indeed, I know enough about magical realism to know that it is a slippery term, much debated and difficult to pigeon-hole. Hence, I shall not belabor the possible biologizing of magical realism, except to note that the term was apparently first employed by Franz Roh, a German art critic, and then applied to literature by the Cuban writer, Alejo Carpentier ("lo real maravilloso"), in the prologue to his novel, *El reino de este mundo* (*The Kingdom of this World*,1949). Subsequently, magical realism--the phenomenon, not simply the phrase--has become especially prominent in the *oeuvre* of such noted Latin American novelists as Mario Vargas Llosa, the early Isabel Allende, Jorge Luis Borges, and, most notably, Gabriel García Márquez. Nor is it limited to writers in Spanish, as can readily be seen by attending to the work of, e.g., Hiruke Mirakami or Salman Rushdie.

There are as many interpretations of magical realism as there are examples. Lost in the hubbub of the trees, however, is this forest: to be magical as opposed to "real," a narrative must somehow depart from our basic understanding of what it means to be biological, and thus, subject to the restrictions of organicity and of the physical world. In this regard, magical realism shares much with fantasy, even–I shall argue–with mystery. This essay focuses upon my own idiosyncratic reading experience, nearly all of it in English; I invite the interested reader to explore possible resonance in other genres, including—not least—that rich vein known as magical realism.

The masters and mistresses of fantasy have long known that whatever else people yearn for; they long to escape our shared biology, to transcend the limitations imposed by life itself. I suspect that when it comes to applying evolutionary biology to literature, the most productive approach is to elucidate the degree to which fiction—despite being, by definition, "made up"—nonetheless adheres to biological insights concerning the reality of human nature (see David P. Barash and Nanelle R. Barash, *Madame Bovary's Ovaries. Delacorte, 2005*). In addition, considerable insight can be gained from seeing literature as a mirror to sociobiological reality (e.g., Jonathan Gottschall's *The Rape of Troy.* Cambridge UP, 2008).

But at the same time, literature is necessarily a matter of the human imagination; accordingly, it should not be surprising that both creators and consumers of fiction often find themselves employing imaginative thought to compensate for their creaturely limits as *Homo sapiens*. After all, despite the oft-cited truism that the human imagination is almost unlimited, it is no less true that we are a particular species of hominid, existing in a particular time, in particular places, with (like any species) our own unavoidable biological baggage.

We are all captive of our personal circumstances—organic, no less than socioeconomic—notably our own bodies and, perhaps to a lesser extent, our minds. "People don't go to the North Pole and fall off icebergs," Chekhov is reputed to have said. "They go to the office, quarrel with their wives, and eat cabbage soup." Much as a prisoner may well fantasize about escaping from his cell, a pervasive theme among those of us who are ostensibly free is to cast off the shackles of our organic, creaturely limitations.

And do what?

Well, if not falling off icebergs, then how about flying: zooming around on a broomstick, like Harry Potter, or drug-assisted, like Carlos Castañeda, or maybe just leaping into the sky, unaided? "It's a bird, it's a plane, it's ...," it's a terrestrial mammal indulging a yearning to transcend gravity and the limitations of being what it really is: a terrestrial mammal. Flight is a hugely compelling theme; it seems that whenever the human imagination is set free, people end up zooming about, above the ground, imitating birds.

So is being invisible. What more compelling fantasy for a blocky, bulky, solidly organic creature such as us than suddenly to become ethereal and unseeable, merged with the air, able to do all sorts of things, yet never called to account, a fly on the wall whenever and wherever one wishes? "Oh that this too, too solid flesh would melt, thaw and resolve itself into a dew," lamented Hamlet. Admittedly, that particular Prince's problems would probably not have been resolved by mere invisibility, just as "the invisible man," in H. G. Wells' eponymous fantasy (1897), discovered that there was a downside to being unseeable.

It is probably no coincidence that the Ring of Power in J. R. R. Tolkien's trilogy, *The Lord of the Rings* (1954-55), rendered its wearer invisible (although it would have been better yet for my purpose if it also conveyed the power of flight). Interestingly, the association of invisibility with power – and, incidentally, with rings – has a long pedigree, going back at least to Plato's *Republic*. Included therein is the story of Gyges, who, we learn,

> was a shepherd in the service of the king of Lydia; there was a great storm, and an earthquake made an opening in the earth at the place where he was feeding his flock. Amazed at the sight, he descended into the opening, ... [and] saw a dead body of stature, as appeared to him, more than human, and having nothing on but a gold ring; this he took from the finger of the dead and reascended. Now the shepherds met together, according to custom, that they might send their monthly report about the flocks to the king; into their assembly he came having the ring on his finger, and as he was sitting among them he chanced to turn the collet of the ring inside his hand,

> when instantly he became invisible to the rest of the company and they began to speak of him as if he were no longer present. He was astonished at this, and again touching the ring he turned the collet outwards and reappeared; he made several trials of the ring, and always with the same result – when he turned the collet inwards he became invisible, when outwards he reappeared. Whereupon he contrived to be chosen one of the messengers who were sent to the court; where as soon as he arrived he seduced the queen, and with her help conspired against the king and slew him, and took the kingdom.

The story of the Ring of Gyges famously combines invisibility with power, although admittedly, it isn't altogether clear how being unseeable also made him sexually irresistible to the queen. In any event, power itself is an immensely attractive component of fantasy. Sometimes it is plain old physical power, the ability to lift heavy objects, crash through walls, demolish one's enemies and delight one's friends. One of the great fantasy novels of all times is *Monkey*, written in seventeenth-century China, and depicting the various adventures – some serious, some slapstick – of (who else?) Monkey, sometimes a monster, sometimes a Buddhist master, sometimes an allegory for human striving, and sometimes, truth to tell, just a plain old monkey. At one point, Monkey obtained a weapon that expanded him, literally:

He at once became a hundred thousand feet high, his head was on a level with the highest mountains, his waist with the ridges, his eye blazed like lightning, his mouth was like a blood-bowl, his teeth like sword-blades. The iron staff in his hand reached up to the thirty-third Heaven, and down to the eighteenth pit of Hell. Tigers, panthers, wolves, and all the evil spirits of the hill and the demons of the seventy-two caves did homage to him in awe and trembling.

The notion of becoming—or seeming—very large (à la Gulliver among the Lilliputians), itself looms large in the imagination. Perhaps it is all too obvious that for centuries people have fantasized about being more powerful, more impressive, and more effective than they actually are. But things that are obvious (e.g., preference for genetic relatives, for revenge, or for the reciprocal aspects of friendship) are no less significant for that. Indeed, they are often especially significant, and the fact that something is taken for granted, rather than diminishing its importance, more often suggests how deep its significance really lies.

Fantasy can also gratify our needs without literally calling for us to change size or shape. One possibility is to be quixotic, to enter into a kind of divine and admirable lunacy whereby we perceive ourselves to be heroic and thus, become so. After he is chastised by a priest for his foolish pursuit of chivalric nonsense, Don Quixote wields his impregnable imagination and defends himself as follows: "I have set injuries and insults straight, righted wrongs, punished arrogance, conquered giants, and trampled on monsters." So there!

Or, if a triumph of the mind alone is insufficient, we might "exceed ourselves" by assuming special powers not otherwise available to other *Homo sapiens*. Like Harry Potter and his fellow wizards in training, we might imagine ourselves possessing special knowledge, familiarity with arcane incantations, the secret handshakes and potions required to rise above our fellows. Or, like the hobbits of Tolkien's trilogy, we may some day discover that despite the fact that we are no more than small and ordinary, we may suddenly be called upon to do great things. As Malvolio intones in *Twelfth Night* (1623), "Some are born great, some achieve greatness, and some have greatness thrust upon them." A perennial theme of fantasy is to be any one of these "greats."

In children's literature in particular it is common for great adventures to be thrust upon one, willy-nilly, without expectation, and often, it seems, without special merit. Think of the children in C. S. Lewis's *Narnia* stories (1950-56), who walk into a wardrobe and find themselves in a magical world in which they are expected to work wonders. Or of Alice, drowsily reading a book on a summer day when she impulsively follows a talking bunny wearing a waist-coat and falls down his rabbit-hole into *Wonderland.* Or Dorothy, transported from Kansas to *Oz*. In all these, the boundary is suddenly erased between that which is real, straight-forward, organically matter-of-fact—our heritage as basically ordinary, "down-to-earth," biological creatures—and something magical, fantastical, and immensely appealing precisely because it transcends our flesh and blood reality with all of its natural limitations.

For instance, consider how many fantasy tales involve interactions with strange and interesting animals, often imaginary concoctions that take zoological reality and mix-and-match so as to produce uncanny creatures that don't really exist, but somehow might: mythological tales of dragons, centaurs or griffins, flying monkeys or other dangerous beasts that must be overcome—such as Grendel in *Beowulf* or Smaug, the dragon, who, as dragons are prone to do, guards a precious treasure in *The Hobbit* (1937). And then there are animals with which we communicate.

Human ancestors are thought to have domesticated dogs about ten thousand years ago (some now claim, 100,000 years ago), followed by cows, horses, goats, sheep, cats, llamas, chickens, and so forth. Our connection with other animals, and not just as predators or prey, thus has an ancient pedigree, so it is consistent with our human creatureliness that so much fantasy literature presents its protagonist as the equivalent of a modern-day horse-whisperer. We need animals, and not just the domesticated ones, no less than they need us. Hugh Lofting's series of *Doctor Dolittle* books (1920-52) are just a small sample. Alice's *Wonderland* is the wonderland of many a grown-up, as well, yearning to talk with the animals: dormice, lizards, turtles and walruses (okay, carpenters and queens, too). Anne McCaffrey's (and now Todd McCaffrey's) popular *Dragonrider* novels (1968-present) touch on a similar yearning, in this case not just the desire to communicate with critters but to experience a profound connection—as companions, mutual care-takers, food providers—of the kind that may well have been especially important in our evolutionary past, and that gratifies some especially fortunate pet-keepers in the present.

Not surprisingly, many yearn to "talk with the animals," to possess the legendary ring of King Solomon, which conferred communicability instead of invisibility. Or instead of seeing the world as others see it, à la Robert Burns, the hope may be to experience the world as other animals do. Thus, in J. M. Coetzee's novel, *Elizabeth Costello* (2003), the eponymous Elizabeth has written a version of *Ulysses* (1922), though from Molly Bloom's perspective. In it, Ms. Costello explains that she intends to go beyond such traditional fiction to imagine, literally, the world as perceived by other species. "If I can think my way into the existence of a [human] being who never existed," she writes, "then I can think my way into the existence of a bat or a chimpanzee or an oyster, any being with whom I share the substrate of life."

This is unlikely. It is certainly the work of a writer to think her way "into the existence of a being who never existed," and such a process is inextricably connected to the fact that we all "share the substrate of life." Strong writers think their way deeply into other human existences, using an intuitive biology that is integral to their writerly strength. But into a "bat or a chimpanzee or an oyster"? Definitely not - for the same reason that human characters *can* be depicted: our shared human "substrate" is that much more real. Nonetheless, Costello's insistence that she can transcend her organic limitations and intuit the private nature of other living things is touching—one of the most widely shared of all human fantasies.

For a final example of fantasy-as-escape-from-biology, let's talk about sex. It is sometimes said that everyone has both a masculine and a feminine side, and yet, the great majority of us are irretrievably dichotomized, signed, sealed, and delivered as either male or female. (For a beautifully realized exception, see Jeffrey Eugenides's *Middlesex*, 2002.) Like Mark Twain's famous comment about the weather, people often talk about androgyny, but they rarely do anything about it. Enter Ursula K. Le Guin and *The Left Hand of Darkness* (1969), a wonderfully wrought tale of the planet Gethen, inhabited by people who aren't so much androgynous as sexless, until they come into "kemmer," a regular period of sexual desire, when they become either male or female, depending on circumstances.

As Le Guin develops the story, Argaven XV, ruler of one of the nations of Gethen, is basically an unsympathetic schmuck, who is kept from being all the more violent and obnoxious by the fact that "he" happens to be pregnant during the events in question. Indeed, Le Guin clearly believes that wars are caused by testosterone poisoning, and her vision of varying, nonexclusive sexual identity is particularly appealing as yet another example of how fantasy can speak to the deeper human need to be—at least on occasion—free of biological constraints.

These come in different forms. They can be physical and overt, imposed by our physiology and anatomy, or they can be more subtle, imposed by behavioral restrictions or by the baggage of one's past. Consider this oft-encountered theme: the unraveling of a secret. The basic pattern here is that as a novel or play unfolds, the reader/audience discovers that underneath a façade of seeming normalcy or even admirable virtue, there lurks something darker. Here's the biological twist: the source of this darkness is not simply the commission of a sin, or a socially inappropriate act, but rather something that threatens to diminish the desirability (often literally, the biological success or "fitness") of the secret-keeper. The reader watches—horrified, fascinated, sympathetic, and sometimes all three—as the truth emerges and its consequences are realized. What unfolds is, in a sense, the biology of secret-keeping: what makes secrets especially troublesome is what also makes it so tempting to keep them secret in the first place, namely, the fact that once revealed, they are hurtful to the secret-keeper's biological interest.

In some cases, as with Ibsen's play, *Ghosts* (1882), there are almost literal ghosts that return to haunt the living. On the tenth anniversary of her husband's death, when the deceased and greatly esteemed Captain Alving is to be memorialized by the opening of an orphanage in his name, his widow feels it necessary to reveal that the Captain wasn't quite the saintly paragon of virtue that everyone had thought. Rather, he was a drunkard, a frequent philanderer, afflicted by an unnamed sexually transmitted disease that was then passed on to his son, Oswald. It also turns out that years earlier, Captain Alving had impregnated their maidservant, who bore a daughter, now employed by Mrs. Alving, and who, as the play progresses, is subjected to sexual advances by Oswald, her half-brother. The orphanage burns to the ground, Captain Alving's illegitimate daughter leaves in anger when her true ancestry is revealed, while Oswald deteriorates into idiocy because of the disease he inherited from his father. A similar message of the troublesome past returned to haunt the living comes through in Ibsen's *The Wild Duck* (1884), in which it is revealed that a young girl—much beloved by her "father"—was actually begotten by someone else, after which the deceived man rejects her, and she kills herself.

A more recent thematic turn comes in Keri Hulme's *The Bone People* (1984), which introduces readers to a New Zealand trinity: Simon, a mysterious mute child who is found on a beach and adopted by Joe, a Maori man who has lost his wife and son and is not only embittered by these losses, but also thwarted when it comes to meaningful family, religious, and cultural ties. The two engage with Kerewin, a strong woman who has her own issues with anger and alienation. Most alienating of all, however, is the reader's discovery of how Joe and Simon are bound together by "a bloody kind of love that has violence as its silent partner." The truth seeps out, and often, it isn't pretty.

Eugene O'Neill's *Long Day's Journey into Night* (1956) presents the equally unpretty unraveling of the Tyrone family, as truths are revealed. Mary Tyrone, the dominating, morphine-addicted mother, famously notes: "The past is the present, isn't it? It's the future, too. We all try to lie out of that, but life won't let us." Neither will biology. We are stuck with our past, not only the stigmata of our evolutionary history, carried behind like tin cans tied to the tail of a tormented dog, but also the personal reality of each person's individual history, behavior that becomes especially consequential insofar as it indicates something about one's character, desirability as a mate, as a parent, as a friend, as an associate, or even as a child worthy of what biologists call, with a certain flinty-eyed detachment, "parental investment."

A case can be made that Joseph Conrad's two best novels are also centered on the question of revealing the truth about past indiscretions, shining a light, eventually, on his protagonist's inner heart of darkness. Thus, when it comes to secrets and the biology thereof, *Heart of Darkness* (1899) isn't simply a story of ivory trading in the Congo; beyond the geographic darkness lies the deeper, metaphoric darkness of the mysterious Mr. Kurtz, full of promise and European enlightenment, who "went native" and engaged in a hidden life of unspeakable atrocities. And *Lord Jim* (1900) tells of a young sailor who, in a moment of panic, abandoned his ship and spent the rest of his life atoning for that act of cowardice. Courage is an appealing trait, thought to be especially attractive to women; its absence, in turn, is likely to diminish one's long-term biological success in projecting genes into the future. Conrad's Jim, not surprisingly, feels diminished by his earlier lack of courage, and his struggle to erase this secret stain drives much of the ensuing novel, concluding in Jim's death.

The unraveling of secrets, with its hidden subtext in which biology lurks in the background, brings us to a literary genre that I haven't yet mentioned: mysteries. From Sherlock Holmes to Lord Peter Wimsey to the abundant and extraordinarily popular novels of Agatha Christie, Earl Stanley Gardner, and, more recently, John Grisham, "detective fiction" has had millions of devotees, with no sign of diminution. Why are who-dunnits so durable? Is there any biology hiding here as well, whose revelation might offer insight?

For one thing, a good mystery story offers the satisfaction of solving an intellectual puzzle, not unlike a crossword or anagram, but with this additional frisson: the reader gets to deal with imagined situations involving at least a simulacrum of real people and real life. Who killed the beautiful ex-wife of the obnoxious, philandering physician? Who robbed Fort Knox? Can anyone stop the shadowy terrorist figure before he blows up Congress? For many, such matters are far more compelling than filling in empty boxes on a page or solving purely theoretical mind-games. If, as seems likely, much of human intelligence evolved in the context of complex social situations, it seems equally likely that human beings have also evolved to employ their intelligence trying to work out precisely puzzles of this sort, with the added payoff that when it comes to murder, robbery, conspiracy, and so forth, the payoffs are higher, and hence, the human interest is greater than when dealing with the more mundane matters that occupy nearly all of our lives, nearly all of the time. Mysteries, in short, give us an opportunity to immerse ourselves, vicariously (and thus safely), in precisely those high-stakes social puzzles likely to hold our attention because they smack of biology: competition, cooperation, success, and failure.

At the same time, since a really good mystery story involves the reader in consequential events in the lives of other people, it may well play into another deep-seated human inclination: our fondness for gossip. In his book, *Grooming, Gossip and the Evolution of Language* (Harvard UP, 1996), British evolutionary psychologist Robin Dunbar suggested that one reason we have language is because we don't groom each other, like other self-respecting primates do. Small groups of monkeys and apes do much of their communicating via touch, not only removing fleas and lice, but at the same time cementing social relationships. According to Dunbar, however, as primitive human beings began to live in social units that were so large that physical contact could no longer maintain social cohesion and permit adequate information sharing, language began to fill the gap between individuals. Gossip is verbal grooming.

Certainly, a highly social species would be not just interested, but highly focused and even fascinated with such questions as "Who is hooking up with whom?", "Who is breaking up with whom?", and of course, more generically, "Who did what to whom?" Such focus and fascination would be especially piqued if the matters at hand involved questions of reputation, status, wealth, love, and particularly, physical injury, life and death. Put this all together and perhaps the popularity of mysteries isn't really such a mystery after all.

Finally, when it comes to the merging of secrets and fantasy – and the biology that most likely underlies their admixture - it makes sense to revisit that old chestnut of psychoanalysis and literary myth, Oedipus. Freud latched onto Sophocles's great tragedy, *Oedipus Rex*, as exemplifying his crackpot notion that somehow, deep inside, boys want to kill their fathers and marry their mothers (and, to a somewhat lesser extent, the converse is true for girls). To that I say "Oedipus shmedipus!" Oedipus himself didn't even have an Oedipus complex; in fact, when told of a prophecy that he would kill daddy and marry mommy he left home specifically so as to foil this fate, whereupon he unknowingly walked right into it. Freud wasn't troubled by this sort of incongruency, however, pointing out that such is the nature of oedipal desire: people are possessed by it without knowing so.

Unquestionably, Oedipus broke a slew of rules, biological as well as social. However unintended, his behavior was shockingly maladaptive, as he recognizes by lamenting these blunders: "To know my mother's bed, to breed children that no man could bear to look at, and murder my father, the father who gave me life!"

My point is that Freud is irrelevant in this case (as in most), although he is often invoked. Take, for instance, Henry Roth's *Call It Sleep* (1934), a lyrical, intense, highly acclaimed tale of the Jewish immigrant experience in the Lower East Side of New York prior to World War I. It is a *Bildungsroman*, focusing on young David Schearl as he navigates the brutal world of New York's turn-of-the-century slums while—even more importantly—steering clear of his violent, hate-filled father, Albert. David's primary refuge and source of love and support is his mother, Genya. Commenting on this triangle, critic Alfred Kazin wrote that "Instead of sentimentalizing the family situation, Roth turned husband, wife, and son into the helpless protagonists of an obvious and uncompromising oedipal situation." But here is something even more obvious: the narrative foundation for the book is the elder Schearl's anguish and fury that years earlier, in "the old country," Genya had a love affair with someone else (worse yet, a Gentile). As a result, young David is not merely unloved, but is resented and even detested by his father, who repeatedly and painfully doubts his paternity. With biology so prominent, who needs Oedipus?

There are other, non-Freudian and far more efficient and biologically less "complex" explanations for the so-called Oedipus complex: notably, the fact that reproduction among close genetic relatives is liable to produce "inbreeding depression," whereby offspring are less viable than those produced by "outbreeding." And of course, all this operates independently of Freudian mythology or other abracadabra.

When incest, which has such resonance in the human mind, occurs in literature, it mainly serves one of two purposes. Some of the most compelling imaginative story-telling, including Gabriel García Márquez's *Cien años de soledad* [*One Hundred Years of Solitude*] (1967), uses seemingly "successful" incestuous relationships to flag their fantastical nature; only in a world of magic can such relationships occur without immense pain. In the real world, it is generally – if paradoxically - a very bad idea to make love with those we love the most. Gritty novels of realism and suffering often take advantage of incest's horror; witness Lawrence Durrell's *The Alexandria Quartet* (1957-60) or Herman Melville's *Pierre* (1852), considered in its day to be scandalous by virtue of its protagonist's obsession with literally loving his sister. By the same token, the horror of young Pecola Breedlove's situation in Toni Morrison's first novel, *The Bluest Eye* (1970), is italicized when she is raped and inseminated by her own father. On a more cheery note, nearly everyone reading *The Color Purple* (1982) breathed a sigh of relief when it was revealed that Celie's rapist wasn't *really*—i.e., genetically—her Daddy.

There exists a nascent movement to examine literature from the perspective of evolutionary psychology, an approach that seems likely to be more productive than postmodernism, Marxism and Freudianism combined. (And maybe we can even throw in queer theory, feminist theory, New Criticism, Old Criticism, and crackpot criticism, too!) Some might argue, however, that fantasy in particular should be excluded from any evolutionary/literary truth-seeking mission, that this sort of literature (involving as it does intentional departures from traditional, earth-bound concerns) should remain a final outpost against biologically-inspired probings. But in fact, an appreciation of fantasy—as with literary illusion generally—is enhanced by recognizing its underlying purpose: allowing the reader to escape, to experience worlds and situations customarily beyond one's capacity or ken. Many *literati*, interestingly, take a dismissive attitude toward the mass appeal of sci fi and fantasy, as well as mysteries, even as they often continue to indulge (in) Freud. Ironically, a biological sensibility on the part of readers might just outfit such works with new protection against ridicule, suggesting that these much-maligned genres speak powerfully to underlying issues concerning human nature itself, albeit usually without the authors knowing what they are up to. These works offer far more than vivid, colorful covers and outrageous plot lines; rather, they show humanity at its most vulnerable, and, on occasion, its most rewarding.

In addition, they bespeak a range of human yearnings whereby magic, fantasy, and mystery provide *Homo sapiens* with both an escape from biological reality and a constant reminder of the underlying "realism" to which our bodies—if not our minds—are unalterably tethered.

CHAPTER 2

The Science Wars in a Long View: Putting the Human in its Place

Imagine you are taking a quiz in the history of modern critical theory. One of the questions asks you to identify the period in which "the science wars" took place. You know that in 1994 Gross and Levitt published *Higher Superstition: The Academic Left and Its Quarrels with Science*. You probably recall that Alan Sokal's celebrated hoax was published in a special issue of *Social Text* designed specifically to answer Gross and Levitt, and you might also remember that Sokal was himself first alerted to the ideas of postmodern science theory by reading Gross and Levitt. So, the middle of the 1990s seems a likely starting point. You are aware, of course, of the debate between C. P. Snow and F. R. Leavis nearly half a century before, but that took place in an entirely different theoretical and cultural context. Leavis was something of a cross between a modernist and an old-fashioned Victorian humanist, and the last feeble remnants of his generation, like the last veterans of World War Two, are now rapidly dying off. Snow's phrase "the two cultures" has entered into the common parlance, but poststructuralism has radically altered the theoretical character of the conflict between the sciences and the humanities. The poststructuralist revolution overwhelmed the old-fashioned humanists some thirty years ago, and the leaders of that revolution have long been firmly established—to use another of Snow's famous phrases—in the corridors of power. So, say we locate the onset of "the science wars" in the mid-nineties. Can we identify a point of conclusion? Some of the contributors on the *Social Text* side of the conflict have declared that there was no war to lose. For instance, as Barbara Herrnstein Smith explains the situation, the cultural constructivists did not mean what Sokal and his cohorts thought they meant. The scientists were just too little versed in rhetoric to grasp the finer shades of ambiguity in postmodern accounts of science. In any case, whether the science wars were worth fighting or not, are they now over? Could either side stand on a carrier and declare, "Mission accomplished"? Having explained that the Sokal brigades were using high explosives to blow up straw men, could the postmodern science theorists declare that they had already explained quite enough, that they find the whole affair tiresome, and that they have more important issues with which to occupy their attention? Or could Sokal and his cohorts affirm that the postmodern theorists, whether or not they used to say the things the scientists thought they were saying, have stopped saying them? Some of the participants in this conflict would no doubt like to declare victory and go home, but off in the distance one often still hears explosions and sees dark plumes of smoke, followed by the wailing of sirens.

For more than thirty years now, beginning with *Sociobiology: The New Synthesis*, E. O. Wilson has displayed a remarkable capacity to incite explosive responses to the claims of science. In *Consilience: The Unity of Knowledge* (1998), Wilson proposed terms for a peace treaty between the two cultures. The terms were simple—*Anschluss*. The humanities would be enfolded within the larger explanatory contexts of evolutionary social science and evolutionary biology. The two cultures would thus become one. And the response from the humanities? Jubilant crowds from the convention hotels at the MLA conference pouring out into the streets to celebrate the New World Order based on the hegemony of science? Hardly. The lessons of appeasement had not been lost on the members of the MLA. Responding to the now incessant cries of crisis in the humanities, Louis Menand declares that whatever else the humanities might need—and many people evidently feel that they badly need something—they do *not* need "consilience." To embrace consilience, Menand explains, would be to make "a bargain with the devil" (14). The humanities have their own distinct provinces, their sacred soil. This they must not give up. Quite the contrary. What they must do instead is "hunt down those disciplines whose subject matter they covet and bring them into their own realm." *Lebensraum*. Eastward lies the course of empire. The vast plains of Russia lie open for the taking. One must only sweep the ground clear of its present unworthy inhabitants—illiterate peasants with names like Gross, Levitt, Sokal, and Wilson. And thus the cycle of violence continues.

At the risk of flunking this particular quiz, I'm going to extend the starting point for the science wars back more than a century. As a battle over curricular turf, the science wars began in 1880, with the exchange between T. H. Huxley and Matthew Arnold. As a battle over what science can tell us about the meaning of life, the conflict goes back further still, but Arnold and Huxley give classic formulations to basic terms in the debate, and they are the first prominent essayists to link issues of metaphysics and cultural history with propositions about the relative standing of science and humane letters within the university. Huxley gives powerful expression to a materialist metaphysic concordant, as he believes, with the revelations of modern science, and he assesses the main phases of Western cultural history from that metaphysical perspective. Huxley's formulations have had little positive influence on the humanities, but among scientists of broad general culture, Huxley remains a living voice. He is cited with respect by E. O. Wilson, and his central contentions have been taken up and reformulated by Steven Weinberg, a Nobel Prize winning physicist, essayist, and prominent participant in the science wars. In responding to Huxley, Arnold affirms an alternative metaphysic and an alternative conception of the human. For nearly a century, Arnold's humanist idealism provided a central guiding light for literary scholars. In some respects, the poststructuralist revolution produced a radical disjunction with the old Arnoldian episteme, but the deepest underlying impulses in Arnold's defense of "humane letters" are still active in postmodern accounts of science and culture. I admire Arnold enough to have written a book on him, but my own metaphysical and epistemological views are more closely aligned with those of Huxley, Weinberg, and Wilson than with those of Arnold and his descendants in the humanities. For the past fifteen years or so, I have been working to establish linkages between literary study and the evolutionary social sciences. In assessing the possibilities for integrating science and the humanities within a single culture, I am thus of the devil's party.

The occasion for Huxley's essay "Science and Culture" was a celebratory address on the founding of a technical college. Huxley approved the provision for a specifically scientific education for a given set of students, and he defended it as an alternative to the emphasis on Greek and Roman literature that prevailed in most higher education at the time. In support of the pedagogical mission of scientific training, Huxley makes three main points about the cultural significance of modern science: (a) science has fundamentally changed our world view—our vision of nature and the place of humankind in nature; (b) nature forms a unitary causal order that can be most adequately accessed using scientific methods; and (c), adopting the ethos of science is an ethical imperative for reasons of both intellectual and social responsibility. In supporting these contentions, Huxley gives a synoptic historical account of four phases of Western civilization, the ancient, the medieval, the Renaissance, and the modern. He attributes to the Renaissance the historical mission of recuperating the culture of ancient Greece and Rome, but he segregates the modern world from all preceding epochs. He argues that modern science enforces a world view that separates the modern period from the Renascence more widely than the Renascence was separated from the middle ages (234-59). As he explains more fully in his essay "On the Advisableness of Improving Natural Knowledge" (1866), before the advent of modern science, humankind had always taken itself "as the standard of comparison, as the centre and measure of the world" (34). He speaks of the animistic fantasies of all primitive peoples and of the ancient Greeks and Romans, and of course he speaks of the spiritualist notions of Christianity and other religions. For Huxley, science in general enforces a materialistic vision of the natural order. Astronomy, first of all, "has filled men's minds with general ideas of a character most foreign to their daily experience" (35). It tells them "that this so vast and seemingly solid earth is but an atom among atoms, whirling no man knows whither, through illimitable space." It "opens up infinite regions where nothing is known . . . but matter and force, operating according to rigid rules" (35). Extending this vision into the range of biology, he affirms that "as the astronomers discover in the earth no centre of the universe, but an eccentric speck, so the naturalists find man to be no centre of the living world, but one amidst endless modifications of life" (37).

In delineating an epochal shift in metaphysical vision, Huxley also identifies a primary source of conflict in the struggle between the humanities and the sciences. In one way and another over the past century, proponents of the humanities have continued to seek to envision humans as the center and measure of the world. Over against that humanistic impulse, the sciences have posed a vision of the world as a vast network of material or physical forces in which the human is but one further link in an unbroken chain of physical causes.

In the terms and concepts Huxley uses for describing cosmology and physiology, he was of course limited to the science available in his own day. The conceptual transformations in these disciplines since Huxley's time have been immense, but Huxley himself would have assimilated these developments with enthusiasm, and they would not have undermined his larger metaphysical vision. We can assess the enduring power of Huxley's vision by comparing it with that of Steven Weinberg, who has been at the forefront of advances in modern particle physics. In *Dreams of a Final Theory* and *Facing Up*, Weinberg outlines the developments in physical knowledge up to the present time, but he also reaffirms the basic principles enunciated by Huxley. He uses one of Huxley's own essays, "On a Piece of Chalk," as the starting point for tracing a sequence of physical causes that leads from common visual perception to the limits of current knowledge in physics. While taking full account of the way in which "principles of symmetry" (*Dreams* 3) have replaced older conceptions of "matter," he follows Huxley in affirming that science "gives us access to the logical order built into nature itself" (*Facing Up* 146). Like Huxley, too, he generalizes from specific discoveries to a larger historical shift in metaphysical vision. He speaks of "the profound cultural effect of the discovery, going back to the work of Newton, that nature is strictly governed by impersonal, mathematical laws" (*Facing Up*). Most importantly, like Huxley, he insists on a certain metaphysical bleakness—a universe of insentient force in which the human occupies a trivial and marginal position:

> Nothing in the last five hundred years has had so great an effect on the human spirit as the discoveries of modern science. . . . We find that the earth on which we live is a speck of matter revolving around a commonplace star, one of billions in a galaxy of stars, which itself is only one of trillions of galaxies. Even more chilling, we ourselves are the end result of a vast sequence of breedings and eatings, the same process that has also produced the clam and the cactus. . .
>
> Some of the old magic has gone out of our view of the role of humanity in the universe, its place being taken by what Matthew Arnold called the "note of sadness." . . .
>
> The human race has had to grow up a good deal in the last five hundred years to confront the fact that we just don't count for much in the grand scheme of things, and the teaching of science as a liberal art helps each of us to grow up as an individual. (*Facing Up* 4-5)

Weinberg's quotation on the "note of sadness" is from Arnold's poem "Dover Beach," in which he describes "The Sea of Faith" as a "melancholy, long, withdrawing roar" (*Poetical* Works 211). In his poems, Arnold spoke for a whole phase of Western culture. He gave touchstone expression to an epochal sense of dismay at the dissolution of a religious vision of the cosmos.

Most of Arnold's poems were written in his younger days. In the last thirty years of his career, he chiefly wrote essays in which he offered remedies for the metaphysical sadness articulated in his poems. It is in this later, consolatory phase of Arnold's thinking that we can locate his essay "Literature and Science." Responding directly to Huxley, Arnold poses the question as to whether the predominance of letters in education ought now to pass to science. To answer this question in favor of literary education, he appeals to Plato's argument that an intelligent man "'will prize those studies that result in his soul getting soberness, righteousness, and wisdom, and will less value the others'" ("L&S," 55). Again invoking Plato, he maintains that the "'fundamental desire'" in "human nature" is the desire for "'good'" ("L&S," 63). In Arnold's thinking, the "good" consists ultimately in the harmonious integration of all the human faculties. He develops a teleological scheme of cultural history, quasi-Hegelian, in which a transcendent force works through history toward a culminating realization of a perfected human condition. In Arnold's cultural theory, literature is the most important medium through which we can achieve this "full humanity."

Superficially, Arnold seems to assimilate Darwinian naturalism. He says that in looking through the findings of modern science, "at last we come to propositions so interesting as Mr. Darwin's famous proposition that 'our ancestor was a hairy quadruped furnished with a tail and pointed ears, probably arboreal in his habits'" ("L&S," 64). With the deft and good-humored wit so characteristic of his essays, Arnold turns this proposition to the advantage of literary studies. On the grounds that mankind has an innate need for the "good," he concludes that our primate ancestor "carried hidden in his nature, apparently, something destined to develop into a necessity for humane letters" ("L&S," 72). By identifying ancient Greek literature as a model for a grand and noble unity of aesthetic perception, he enables himself to draw the still more dramatic inference "that our hairy ancestor carried in his nature, also, a necessity for Greek."

Arnold's manner is charming, and his logic is beguiling, but his facile blending of Plato and Darwin is deceptive. In Darwin's theory of natural selection, there is no transcendent teleological force driving toward some culminating historical realization of psychological and cultural harmony. The driving force in human evolution is the mechanical process of natural selection. The regulative principle that has shaped human nature, as it has shaped the nature of every other species, is inclusive fitness—the transmission of genes. Inclusive fitness has designed human nature in such a way that conflicts of interest are integral and ineradicable. Relations between men and women, parents and children, siblings and other kin, and individuals within a social group—all these relations involve tensions between reciprocal benefits and competing interests. Humans have evolved distinctive capacities for cooperation within groups, but the larger context for the evolution of cooperation within groups is the conflict between competing groups. Within the Darwinian conception of human evolution, there is no transcendent teleological process guiding human cultural history, and there is no transcendent aesthetic and ethical order to which the human mind, through culture, can gain access.

Huxley and Arnold were friends, and the tone of their references to one another is genial. They nonetheless differ profoundly in metaphysical and historical vision, and that difference plays itself out in the subsequent history of the debate over the sciences and the humanities. Huxley announces a radical break in the modern world view. Arnold, in contrast, gives a strong emphasis to continuity in the cultural imagination of European civilization. Huxley had spoken with scathing contempt of medieval superstition. In reply, Arnold grants that the medieval cosmology is obsolete, but he defends the quality of medieval education on ethical and aesthetic grounds. He concedes that we can no longer uphold traditional Christian beliefs, but he argues that we have no need of those beliefs. We can, he says, look instead to poetry and the other arts to satisfy the emotional and imaginative needs that religion once satisfied. This is a radical proposition. Arnold argues that the real and effective part of religion has always been its unconscious poetry. In the future, he thinks, poetry itself, detached from religious belief, will fulfill all the psychological and moral needs once fulfilled by religion. "The strongest part of our religion to-day is its unconscious poetry. The future of poetry is immense, because in conscious poetry, where it is worthy of its high destiny, our race, as time goes on, will find an ever surer and surer stay" ("On Poetry" 63).

From our current vantage point, Arnold's proclamation on the future of poetry seems quaint. Almost no one at the present time would invest poetry with this weighty mission. For nearly a century, though, Arnold's belief in the mission of literature had an immense influence in academic literary study and in the wider culture. Until about 1970, the majority of literary scholars could reasonably have been described as "humanists" who shared some important part of Arnold's literary idealism. A humanist in this sense is a scholar who invests literary subjects with an almost sacred value. Such scholars believe that literary works give human beings access to a spiritual realm in which some ultimate harmony or resolution can be glimpsed. Arnold's own humanism mingles Platonic transcendence and Wordsworthian piety with a frank admiration for creative literary genius. He describes great writers as "gifted men, alive and active with extraordinary power at an unusual number of points," and he says that their works "have a fortifying, and elevating, and quickening, and suggestive power" ("L&S," 68). The commonly accepted views on such matters have now changed so dramatically that younger scholars might find it hard to credit the fervor with which they were once held. In the mid-century period, Arnoldian idealism informed the New Critics' belief in the poem as a verbal icon and Northrop Frye's Romantic and mystical belief in an ultimate order of literary words equivalent to the mind of God—"the anagogic phase." Arnoldian idealism entered in an attenuated form into Lionel Trilling's concept of literary culture as an apex of "the liberal imagination," and it worked its way also into F. R. Leavis' defense of literary culture over against the claims of science.

The exchange between C. P. Snow and F. R. Leavis is in some ways more important for its symptomatic value than for its substantive intellectual content. Snow speaks of physical science as "the most beautiful and wonderful collective work of the mind of man" (*The Two Cultures* 14), but he says nothing of its metaphysical character and very little of its epistemological character. He feels that humanists should know more about physical science and that scientists should read more novels, but he does not seem to register that there are any ultimate questions of meaning at stake. Snow's vision is essentially utilitarian. He advocates more science education chiefly on the grounds that it will produce greater physical comforts for the mass of society. He is sympathetic to the Soviet model of education, and he seems to feel that shared norms of material comfort will eventually heal the ideological wounds of the modern world. In his essay responding to Snow, Leavis gave expression to an aesthetic and moral revulsion that shocked a good many people. Arnold wished to embody in his own prose the genial urbanity of "culture." Leavis aims instead at evoking and denouncing the spiritual emptiness he detects in Snow's techno-managerial perspective on society.

Leavis is no Platonist. His ideal of poetry filters itself through D. H. Lawrence's passionate individualism, but he nonetheless invests that ideal with the power to replace religion and provide spiritual meaning in the modern world. In opposition to Snow's vision of a social collective managed by a bureaucratic elite, Leavis appeals to the Laurentian maxim that "'nothing matters but life'" ("Significance" 53). He concurs with Lawrence's belief that "only in living individuals is life there, and individual lives cannot be aggregated or equated or dealt with quantitatively in any way" ("Significance" 53-54). Leavis' formulation is typical of a view very wide spread in humanistic thinking—a kind of dualism that separates the world into two parts: a physical natural order that can be known by science and a purely qualitative, subjective human realm that can be accessed only through discursive modes. Leavis presents this subjective realm as an answer to questions about ultimate meaning. "In coming to terms with great literature we discover what at bottom we really believe. What for—what ultimately for? What do men live by?—the questions work and tell at what I can only call a religious depth of thought and feeling" ("Significance" 56). Leavis regards the challenges of rapid technological change chiefly as a threat, and he argues that literary education is a necessary means for meeting that threat. Echoing one of Arnold's phrases, he speaks of having recourse to "our full humanity" ("Significance" 60).

I've said that the essays by both Snow and Leavis seem more important for their symptomatic value than for their actual intellectual content. With respect to Leavis, what I have in mind is the vacuity of the rhetoric through which he seeks to evoke the "full humanity" supposedly to be found in literary training:

> [Mankind will need] a basic living deference towards that to which, opening as it does into the unknown and itself unmeasurable, we know we belong. . . . What we need, and shall continue to need not less, is something with the livingness of the deepest vital instinct; as intelligence, a power—rooted, strong in experience, and supremely human—of creative response to the new challenges of time. ("Significance" 60-61)

This is a desperate sort of rhetoric—straining for intensity of effect, hyperbolic, disjointed, and vacant of substantive propositions. At the distance of half a century, looking back, we can reasonably suggest that Leavis' proclamations represent something like a spasmodic last gasp for old-fashioned literary humanism. The strained and vacant intensity of the rhetoric by which he claims a central place for a literary education goes a long way toward explaining the ultimate collapse of the Arnoldian rationale that sustained the humanities through the first half of the twentieth century. It simply would not wash.

Over the past thirty years or so, two major revolutions have taken place across the disciplines—deconstruction, with its radiations into Foucauldian discourse theory, and sociobiology, with its radiations into evolutionary psychology and behavioral ecology. The deconstructive revolution jettisoned Arnoldian humanism and took on a completely new set of authorities, mostly French. In some ways, the introduction of poststructuralist or postmodernist thinking in the humanities represents a simple reversal in attitudes and concepts that had previously characterized literary training. In place of Platonic idealism and the appeal to some ultimate harmony in culture, deconstruction and Foucauldian discourse theory tend toward metaphysical nihilism and subversive ideology. In place of the appeal to the creative power of gifted individuals, postmodernism transforms the individual into a passive vessel for the circulation of cultural energies. Nonetheless, in two crucial respects, the deconstructive revolution retains continuity with Arnoldian humanism—in its emphasis on verbal culture, and in placing the claims of ethical values over the claims of objective knowledge. Huxley affirms that science bids us "seek for truth not among words, but among things" (*Science* 150). Postmodern thinking, in contrast, locates ultimate epistemic authority in words—in discourse, in language or semiosis. It systematically deprecates the possibility of objective, empirical knowledge of a real, physical world that exists independently of any human discourse, and it is, in this respect, still qualitative and verbal in orientation. The appeal to values in the deconstructive dispensation is an appeal against established structures of social power. In its programmatic repudiation of existing power structures, postmodern thinking in the humanities seems virtually to invert the conservative tendencies in Arnoldian cultural idealism, but it is continuous with Arnoldian cultural idealism in basing its claim to cultural authority on its claim for wisdom and justice—for representing an enlightened ideological consciousness.

In the same period that poststructuralism was establishing itself in departments of literature, an evolutionary view of human behavior was establishing itself in the social sciences. This revolution is not yet so complete as the poststructuralist revolution, but it seems virtually guaranteed a permanent public success. It is grounded in the empirical reality of human evolution, and it is wedded to an empirical methodology that provides it with the means for the correction of error and the cumulative development of knowledge. One such correction has been taking place within just the past few years. In its early days, sociobiology invoked the idea of "selfish genes" and discounted the possibility of "altruistic" traits selected for the advantage of groups. More recently, the idea of "group selection" or "multi-level selection"—the selection of traits designed to facilitate group living—has gradually reasserted itself as a necessary element in the larger picture of human social evolution. In parallel with that development, leading thinkers in evolutionary social science are now recognizing that human beings have evolved a unique capacity for sustaining and developing culture. I think it safe to predict that within evolutionary social science, the next major phase of development will involve a new understanding of what is known as "gene-culture co-evolution"—the interdependence of genetic evolution and cultural development in humans.

Since the time of Huxley and Arnold, seemingly disparate ideological values in the humanities have converged in a common desire to maintain human experience as "the centre and measure of the world." From the perspective shared by Huxley, Weinberg, and Wilson, that desire entails a false notion of where humans stand in the general scheme of things—a false cosmology, and a false understanding of human evolution. The idealizing sentiments of Arnoldian humanism have now largely faded from sight, and they are not likely to return. The more recent strategy in the humanities has been not to exalt the human but rather to delegitimize the idea of objective scientific knowledge while simultaneously elevating "discourse" to an ultimate ontological category. The cultural study of science from a postmodern perspective deprecates the idea of human nature, but discourse is itself a specifically and distinctively human function. No other species has developed a language sufficiently complex to articulate propositions about the ontological primacy of language. By subsuming the knowledge of nature within the philosophy of "discourse," postmodern science theory indirectly, from the back door, reaffirms the centrality of the human as the measure of all things.

From a consilient perspective, arts and letters—the subjects of the humanities—are encompassed within more elementary domains of knowledge, within psychology, anthropology, and evolutionary biology. The human sciences and life sciences are themselves encompassed within the still more elementary causal domains of chemistry and physics. Literary scholars who accept this consilient conception of their field would not claim that the theory of discourse has an ultimate epistemic authority. They could, however, identify the concepts and concerns that are particular to literature and other humanistic subjects; they could integrate those concepts with broader, deeper causal principles from other domains; and they could in this way gain for their field an empirical validity and a power of progressive development greater than it has ever had before.

Scholars in the humanities have primary access to artifacts of cultural imagination. That access could enable them to play a vital role in developing an adequate evolutionary understanding of culture. We need to probe the relations between the evolved cognitive and behavioral dispositions of human nature and the phenomenal properties of literary texts: representations, subjective affects, aesthetic qualities, and formal structures. We also need to locate the regularities of human nature within the particular contexts of literary and cultural history. To contribute effectively to research of this kind, scholars in the humanities would need to expand their base of knowledge and incorporate new skills, including, for some of us, skills of empirical methodology and quantitative reduction. Scholars who adopt these strategies would be joining a collective scientific effort to put the human in its place. That effort need involve no repudiation of their own humanity, but it would probably suggest new ways to envision that humanity. Humans are imaginative animals. New ways of seeing are new ways of being. In adapting to the changing environment of knowledge, we shall probably discover new forms for what we regard as our "full" humanity.

Works Cited

Arnold, Matthew. *Culture and Anarchy*. Ann Arbor: University of Michigan Press, 1965. Vol. 5 of *The Complete Prose Works of Matthew Arnold*. Ed. R. H. Super. 11 vols. 1960-77.

---. "Literature and Science." *Philistinism in England and America*. Ann Arbor: U of Michigan P, 1974. Vol. 10 of *The Complete Prose Works of Matthew Arnold*. Ed. R. H. Super. 11 vols. 1960-77. 53-73.

---. "On Poetry." *The Complete Prose Works of Matthew Arnold*. Ed. R. H. Super. Vol. 9. Ann Arbor: U of Michigan P, 1974. 60-70.

---. *Poetical Works*. Ed. C. B. Tinker and H. F. Lowry. London: Oxford UP, 1950.

Herrnstein Smith, Barbara. *Scandalous Knowledge: Science, Truth, and the Human*. Edinburgh: Edinburgh University Press, 2006.

Huxley, Thomas Henry. *Method and Results*. New York: D. Appleton, 1899. Vol. 1 of Collected Essays. 9 vols. 1896-1902.

---. *Science and Education*. New York: D. Appleton, 1898. Vol. 3 of Collected Essays. 9 vols. 1896-1902.

Leavis, F. R. "Two Cultures? The Significance of Lord Snow." *Nor Shall My Sword: Discourses on Pluralism, Compassion, and Social Hope*. London: Chatto and Windus, 1972. 53-73.

Menand, Louis. "Dangers Within and Without." *Profession* 2005. Ed. Rosemary G. Feal. New York: Modern Language Association of America, 2005.

Snow, C. P. *The Two Cultures*. Cambridge: Cambridge UP, 1993.

Weinberg, Steven. *Dreams of a Final Theory: The Search for the Fundamental Laws of Nature*. New York: Pantheon, 1992.

---. "Sokal's Hoax." *Facing Up: Science and Its Cultural Adversaries*. Cambridge: Harvard UP, 2001. 146-99.

CHAPTER 3

Case Study: The Trickster Genre and the Free Rider Problem

For the bulk of their evolution, *H. sapiens* and perhaps earlier hominid species have lived as hunter-gatherers in small groups of roughly 25 to 100 individuals (Lee and DeVore; Kelly) dependent upon one another for food (e.g., Cashdan; Lee; Rassmussen, "Netsilik"; Tonkinson; Wiessner), health-care (Sugiyama and Chacon), and defense (LeBlanc and Register). It follows that a major selection pressure on ancestral humans was their own kind (Alexander; Byrne and Whiten; Flinn, Geary and Ward): both in-group and out-group individuals could potentially further or foil one's survival and/or reproductive goals. Thus, our ancestors "needed to construct . . . a social map of the persons, relationships, motives, interactions, emotions, and intentions that made up their social world" (Cosmides and Tooby 163). Leda Cosmides and John Tooby argue that this hypothesis "must be substantiated by experimental evidence, as well as by converging lines of empirical support drawn from related fields such as neuroscience, linguistics, and anthropology" (163). I propose that folklore be added to this evidence. There is a pronounced correspondence between the themes of world folklore and the challenges associated with navigating the social world (Scalise Sugiyama, "Folklore"; Scalise Sugiyama, "Reverse-Engineering"). By enabling us to observe a wide range of human actions, personalities, relationships, emotions, and goals, narrative offers a low-cost means of enhancing our ability to predict and manipulate the behavior of others (Scalise Sugiyama "On the Origins"; "Reverse-Engineering").

The prediction and manipulation of behavior is part and parcel of a larger human capacity. Our species' ecological niche—the way we make a living—is the cognitive niche (Tooby and DeVore): what distinguishes humans from other animals is our highly developed ability to make cognitive models of our environment. John Tooby and Irven DeVore call this ability *causal* or *instrumental intelligence*, which they define as the ability to create and maintain cause-and-effect models of the world as guides for prejudging which courses of action will lead to which results. Causal reasoning enabled our ancestors to develop complex resource extraction techniques, such as locating game by tracking, capturing elusive animals through the use of snares, traps, or lures, and neutralizing plant toxins by leeching or cooking. These complex techniques often entail modeling the social as well as the physical environment: game drives, for example, require coordinated human action and division of labor. In the case of Mbuti net hunting, men enclose an area of jungle with a net "fence" and women act as beaters, driving game into the enclosure (Turnbull). The Washo used a similar technique, using brush instead of nets: some group members made the enclosure, others drove antelope toward it, and others shot the antelope once they were inside (Downs). The use of game drives, then, entails modeling (1) applications of physical materials (e.g., brush can be used as an enclosure, plant fiber can be woven into a net, rock can be sharpened to pierce flesh), (2) animal behavior (e.g., loud noises cause certain animals to break cover and run in the opposite direction of the noise), and (3) human behavior (e.g., groups of humans can cause animals to run in a given direction by chasing them).

As the use of conspecifics to effect a plan suggests, cognitive modeling has applications beyond subsistence: it enables one to model possible actions--and reactions--of other human beings. If, for example, a group of strangers is encroaching on your territory, you can model different courses of action and possible outcomes in order to decide how to act. What might happen if you attack them? What might happen if you befriend them? What might happen if you do nothing? Nicholas Humphrey refers to this social application of modeling as a special case of "forward planning" ("Social Function" 19) and likens it to a chess game: any one move may provoke a variety of countermoves, to each of which the player must have a response planned (see also Alexander).

As it turns out, good chess players have a large repertoire of board configurations in their memory, which gives them the advantage over less experienced players when calculating the potential ramifications of a particular move (Groot). By the same token, the more experience one has at observing human behavior, the better one will be at calculating the potential ramifications of a particular action. This is the essence of Humphrey's argument regarding what he calls "doing psychology" (*Consciousness Regained*). In this seminal work, Humphrey argues that, across human evolution, (1) individuals who were better at predicting the responses of conspecifics had higher survival rates than those who were less adept at this task, and (2) accurate predictions of behavior were positively correlated with depth and breadth of personal experience. He concludes that "the extension of inner experience should itself be a biologically adaptive trait in human beings" (69). Richard Byrne makes a similar observation in the context of primate intelligence: noting that at least some chimpanzees appear to be capable of perceiving and applying resemblances between a past experience and a present problem, he argues that this type of problem-solving involves "having plenty of similar past experiences to draw upon, and an efficient memory" (85).

This is precisely the role that narrative plays in problem-solving: it extends "inner experience" by modeling different circumstances, actions taken in response to those circumstances, and the outcomes of those actions (Scalise Sugiyama, "Folklore"; Scalise Sugiyama, "Reverse-Engineering"). As Steven Pinker puts it, "Characters in a fictitious world do exactly what our intelligence allows us to do in the real world. We watch what happens and mentally take notes on the outcomes of the strategies and tactics they use in pursuing their goals" (541). In short, narrative expands our repertoire of life experiences, which we can reference and apply to the solution of present problems, including predicting and manipulating the behavior of others. In the language of Cosmides and Tooby's argument, narrative maps possible relationships, motives, interactions, emotions, and intentions that we may encounter in our social world ("Cognitive Adaptations"163).

One behavior-prediction problem that recurrently beset our ancestors is deceit. Group living inevitably leads to conflict, as individuals pursue goals that are at odds with those of their fellows. Through the use of deceit, one individual may cause another to engage in behavior that furthers the former's goals and/or thwarts the latter's. A subset of deceit is the free rider problem. To reduce certain risks and increase the odds of attaining certain goals, humans form alliances that involve delayed reciprocal exchange: one partner does a favor for the other with the understanding that the favor will be reciprocated at some future date. There is always a chance, however, that one's partner will cheat—that one will be tricked into doing a favor for someone who doesn't plan on reciprocating.

Although humans have cognitive adaptations dedicated to the problem of deceit (e.g., Axelrod, *The Evolution of Cooperation*; Axelrod and Hamilton; Byrne and Whiten; Whiten and Byrne), these adaptations only address facets of the problem that recur across time and cultures. Cognitive mechanisms cannot evolve to anticipate the particular stratagems that particular individuals in particular environments may invent to trick their fellows. Novel information must be acquired either from experience or from conspecifics. Acquiring novel information socially is often much more efficient and much less risky than acquiring knowledge at first hand, and humans are not the only animals to do so. Red-winged blackbirds, for example, learn the palatability of foods by observing what other red-winged blackbirds eat (Byrne). What is striking about humans is the degree to which we learn socially: compared to other animals, we acquire a wider variety and greater quantity of information from conspecifics.

A few examples of the software requisite to human cultural transmission will suffice to make the point that humans are designed to acquire information from others, and to illustrate the complexity of this design. Michael Tomasello identifies a number of cognitive components of human cultural transmission that appear to be absent or much less developed in other species. One is the ratchet effect: "faithful social transmission that can work as a ratchet to prevent slippage backward—so that the newly invented artifact or practice preserves its new and improved form at least somewhat faithfully until a further modification or improvement comes along" (Tomasello 5). While many non-human primates are capable of making innovations, they appear to lack the cognitive abilities requisite to copying them (Kummer and Goodall).

A second critical component of human cultural transmission is intentional reasoning: the understanding that things that behave (e.g., human and non-human animals) have goals, and that these goals influence their behavior (Baron-Cohen; Barrett; Dennett). This ability enables us to explain the behavior of others in terms of the goal it is directed toward. This in turn enables us to visualize how a person might use a given artifact—to "understand the intentional significance of the tool use or symbolic practice" (Tomasello 6).

Yet another critical component of human cultural transmission is what Tomasello refers to as "joint attention." Late in the first year of life, infants develop the ability to look where an adult is pointing (gaze following), check to see if the adult is looking at the same object they are looking at (checking attention), act on objects in the same way adults are acting on them (imitative learning), and direct adult attention to objects by pointing at or holding up an object. In a study involving 24 children, these components emerged in most cases between nine and twelve months, and in distinct order: 20 out of 24 infants first mastered tasks that required sharing or checking adult attention, then mastered tasks that required following an adult's gaze, and then mastered tasks that required directing adult attention to an object (Carpenter, Nagell and Tomasello).

Given, then, that humans are designed to acquire certain kinds of information from conspecifics, an important question follows: What kinds of information are humans designed to acquire from conspecifics? This paper provides a partial answer to that question. As Tooby and Cosmides explain, the mind contains mechanisms—known as *motivational systems*--designed to guide attention to environmental stimuli that recurrently impacted fitness. For example, one way the mind guides behavior is by causing us to experience pleasure when we do something that increases our chances of surviving or reproducing (e.g., eating, sleeping, sex). Aesthetic preferences are another example: we find a given object, activity, or phenomenon attractive because it exhibits cues that, in ancestral environments, signaled that it would be advantageous to pay attention to it. In short, the mind is designed to notice and input information that, in ancestral environments, tended to increase the individual's chances of surviving and/or reproducing.

Motivational and attentional systems cast folklore themes in a new light. For tens of thousands of years, narrative has been oral (Scalise Sugiyama, "Food, Foragers"). This means that over the course of the development of the narrative faculty, story content was constrained by the limitations of memory—specifically, by the kinds of information human memory is designed to input and store. In effect, oral traditions are maps of information regions that were instrumental to survival and/or reproduction in ancestral environments. Two conclusions follow: (1) the themes that recur cross-culturally in oral traditions correspond to domains of information that the mind is designed to attend to, input, and store; and (2) the information contained in oral traditions increased our ancestors' chances of surviving and/or reproducing.

This paper deals with one of those themes—the trickster—and the information domain to which it corresponds—deceit in social relations. Specifically, I argue that the universality of the trickster theme is one line of evidence in support of the hypothesis that humans have evolved mechanisms dedicated to social exchange. The facet of this problem addressed by oral narrative is information storage—specifically, storage of information relevant to effecting, detecting, preventing, thwarting, and/or punishing deceit. The trickster genre contains a multitude of free-rider scenarios: possible goals, possible tricks used to attain those goals, and the consequences of those tricks for the victim, the trickster, or both. The remainder of this paper will explore various components the free rider problem, and how trickster tales model them.

Evidence that humans have evolved an integrated collection of cognitive mechanisms dedicated to navigating their social world has been steadily accumulating over the last two decades. Perhaps the best known of these posited mechanisms is the theory of mind module (ToMM), which emerges predictably around age four-and-a-half (the absence of ToMM is diagnostic of autism; see Baron-Cohen; Baron-Cohen et al.; Frith; Leslie "The Theory," "ToMM"). "Theory of mind" refers to the ability to understand that other people have thoughts and beliefs ("mental representations"), which in turn enables us to intuit and predict the beliefs of others. Theory of mind also enables us to lie—that is, to understand that two people can have different beliefs about the same object. Obviously, this ability is integral to the practice of deceit in social relations.

A fundamental problem of group living is conflict of interest. Group living inevitably entails competition for access to resources (e.g. food, water, mates). Resources are finite, a situation exacerbated by the fact that each individual would like to monopolize the best of those resources—for example, the ripest and most prolific berry patch, the most productive and reliable fishing hole, the most fertile and fit mate. These conditions favor bullying, whereby the biggest or most formidable individual gets the choicest goods by intimidating others. However, the meek are not without recourse: individuals can form alliances to defend themselves against the predations of other group members.

This is where reasoning about social exchange enters the picture. An alliance is a social contract, the terms of which are "I'll help you if you help me." In other words, one individual does a favor for another with the understanding that the favor will be reciprocated at some undetermined time in the future. As noted above, there is always a chance that the recipient of the favor will not reciprocate: this is known as the "free rider" problem. In order for cooperation to be evolutionarily stable within a population, the free rider problem must be solved. To this end, several conditions must be met: individuals must (1) encounter the same individuals repeatedly, (2) recognize individuals they have met before, (3) remember whether an individual has reciprocated them or not; and (4) prevent cheaters from taking a benefit without paying the cost or exclude cheaters from future exchange interactions (Axelrod; Axelrod and Hamilton). In other words, individuals must be able to detect and punish cheaters.

Over the last twenty-five years, through an exhaustive series of experiments, Cosmides and Tooby and their associates have amassed a compelling array of evidence that the human mind contains algorithms for guiding reasoning about social contracts (for a review, see Cosmides and Tooby, "Neurocognitive Adaptations"). Many of these experiments utilize the Wason-selection task, in which subjects are asked to identify violations of the conditional rule, "If P, then Q." When the content of the rule is abstract or descriptive (e.g., "If a person eats hot chili peppers, then he will drink a cold beer"), only 5-30% of subjects respond correctly. However, when the rule is expressed as a social contract (e.g., "If you give me your watch, I'll give you $20"), 65-80% of subjects give the correct response, *P* and *not-Q* (Cosmides and Tooby, "Cognitive Adaptations"). Replication of these results among Amazonian foragers indicates that this reasoning capacity is universal and not the result of cultural processes (Sugiyama et al., "Cross-Cultural Evidence"). Administration of the Wason-selection task to a patient with bilateral limbic system damage supports this conclusion (Stone et al.). The patient, R.M., was given two types of reasoning problems that take the form "If P, then Q": social contract rules ("If you take the benefit, then you must pay the cost") and precaution rules ("If you engage in hazardous activity X, then you must take precaution Y"). R.M. performed significantly worse on social contract reasoning than on precaution reasoning, which suggests that the human capacity for reasoning about social exchange can be impaired, and thus that it is a specialized cognitive mechanism distinct from other kinds of social reasoning mechanisms (Stone et al.). A study of schizophrenic patients supports the hypothesis that the faculty for reasoning about cheaters is separate from other reasoning faculties: whereas patients' general logical reasoning abilities were impaired, their cheater-detection abilities were not (Malijkovic).

Merely detecting free riding, however, is not sufficient to prevent it. Tellingly, evidence from experimental economics shows that humans are willing to incur costs to punish free riders (Dawes, Orbell and Van de Kragt; Fehr and Gachter; Ostrom, Waler and Gardner; Sato; Yamagishi), and that overall levels of cooperation increase when the likelihood that free riding will be punished is sufficiently high (Fehr and Gachter; Kurzban, McCabe, Smith and Wilson). Punitive sentiments toward cheaters are posited to be the product of a motivational subsystem dedicated to preventing "free rider designs from having higher fitness than cooperator designs" (Price, Cosmides and Tooby 228). This hypothesis is supported by survey data, which show that willingness to participate in a collective action is a strong predictor of punitive sentiment toward free riders (Price, Cosmides and Tooby).

As noted in the previous section, adaptations for reasoning about social exchange can only address facets of the free rider problem that recur across time and cultures; they cannot anticipate particular stratagems that might be used, and they cannot prevent cheating. In the next section, we will see how narrative provides a means of surmounting these obstacles.

The trickster character appears cross-culturally in a variety of forms (Bierhorst; Radin). The trickster is sometimes a human being, sometimes an animal but, as Stith Thompson observes, the "distinction is never very clear," and in "the most human of these tales . . . the animal nature of the trickster seems always in the background of the narrative" (*Tales* xviii). Some of the better-known tricksters are Anansi, Brer Rabbit, Coyote, Reynard the Fox, leprechauns, and fairy folk, but there are many others. In North America alone, for example, there are coyote, mink, blue jay, spider, beaver, porcupine, turtle, moose, and rabbit tricksters (Bierhorst; Erdoes and Ortiz; Norman; Rasmussen, "Observations"; Radin; Thompson *Tales*). When it comes to cheating, the trickster is a violator of many rules, not just of social contracts; consequently, trickster tales contain information about many different kinds of deceit in social relations. Free riding is a decided part of this repertoire, and is the focus of this discussion. It should be noted that, in the context of social exchange, the terms "free riding" and "cheating" refer to taking a benefit without paying a cost, or accepting a favor without reciprocating.

Indigenous storytellers, folklorists, and anthropologists commonly observe that trickster and other traditional tales function, in part, to delimit behavioral boundaries by providing examples of appropriate and inappropriate behavior and their consequences. Barre Toelken, for example, asked a Navajo storyteller why Coyote stories are told. The response was, "'If my children hear the stories, they will grow up to be good people; if they don't hear them, they will turn out to be bad'" (101-102). One of Morris Opler's informants compared Jicarilla mythology to a Testament, an observation which Opler seconded: "Jicarilla mythology is truly the codification of the beliefs and mores of the group. There are few matters of conduct which the Jicarilla do not refer back to this body of lore. It is thought of as the proper guide to action, positive and negative" (viii). Grenville Goodwin makes a similar observation, noting that among White Mountain Apache, Coyote tales in particular often have a moral, and that "parents or grandparents after telling these to children, often gave a short lecture to the youngsters: 'Don't do like Coyote did in the story. He did a lot of bad things for us long ago such as marrying his daughter and stealing!'" (ix). Brian Street argues that trickster tales "can be seen as moral examples re-affirming the rules of society; or rather they serve as a model for these rules, demonstrating what happens if the prescriptions laid down by society are not observed" (85).

In cognitive terms, these "guides," "examples," and "morals" are cause-and-effect models of the human social world that can be used to predict which courses of action will lead to which results. In the case of trickster tales, these models address such questions as, In what ways might other people try to cheat me? What might happen to me if I try to cheat someone? Is it ever acceptable to cheat and, if so, when? The different schemes and scenarios presented in trickster tales enable the listener to "observe" a wide variety of stratagems and a wide variety of responses to free riding (both the characters' and the audience's), thereby learning what behavior is allowed in their society and what behavior is not.

Trickster tales contain two broad classes of information relevant to the free rider problem: ways of cheating and consequences of cheating. As noted above, solution of the free rider problem requires four conditions: (1) regularly encountering the same individuals, (2) recognizing individuals, (3) remembering whether a given individual has reciprocated, and (4) punishing non-reciprocators. The stratagems used by tricksters typically foil one or more of the mechanisms that evolved to enable these conditions. Thus, trickster tales contain information that is useful for recognizing a potential scam (or for planning one) and/or for discouraging would-be cheaters (or for avoiding punishment).

An example of the former is the Cree tale, "Trickster and the Shut-Eye Dancers" (Norman). In this story, Trickster and Fox finally manage to catch three ducks after more than twenty-four hours of effort. Fox wants to eat them immediately, but Trickster says they will taste better with some other-side-of-the-marsh grass sprinkled on them. While Fox travels across the marsh to get the grass, Trickster roasts the ducks and eats them all. The implied contract in this scenario is, "If you help me catch ducks, I will share them with you." The obvious lesson of the story is to avoid giving your partner the opportunity to break his/her word. (From the perspective of a free rider, the story presents information regarding a ruse that might be employed to trick someone out of food.) Fox creates this opportunity by leaving the ducks with Trickster and going to the other side of the marsh.

One of the troublesome facts folklorists have wrestled with over the years is that many tricksters are also culture heroes. As Thompson notes, the trickster may simultaneously be the bringer "of culture and light to his people" and be "the incarnation of greediness, lust, cruelty, and stupidity" (*Tales* xviii). An information-transmission perspective easily resolves this paradox. Cheating can be good or bad, depending on context and perspective. For example, a trick that enables one to acquire an indispensable and otherwise unattainable resource or to weaken an aggressive and formidable enemy can be a very useful tool. Accordingly, trickster tales implicitly delineate a culture's tacit rules regarding when and whom it is acceptable to cheat. Generally speaking, cheating enemies or out-group members is condoned, and cheating in-group members (family, friends, band members) is not.

For example, in Mandan-Hidatsa culture, Coyote's "traits are found desirable to the Indian in their conflict with enemy tribes and with the antagonistic forces of nature. . . . he is sly and tricky, a good thief, capable of providing for himself when provisions fail" (Beckwith xvii). Similarly, when a story is told among the Skidi Pawnee "in which Coyote, or some culture-hero overcomes his enemy, the teller thereby indicates his desire that he also may be equally successful in some venture which he has in hand" (Dorsey xxii). The Salishan tell a story that exhibits a similar attitude, wherein Coyote steals fire for the people. Coyote recruits the help of good runners, such as Fox, Wolf, and Antelope, and travels to the home of the Fire people, where he announces that they have come to dance, play, and gamble. The Fire people make preparations for a dance, and Coyote makes himself a headdress out of pitchy yellow-pine shavings and dry cedar bark. They then begin to dance around the low fire, and

Coyote complains that he can't see so that they will make it bigger. He keeps this up until eventually the fire is blazing, at which point he

> ...danced about wildly until his head-dress took fire. He pretended to be afraid, and requested the Fire people to put it out. They warned him not to dance so close to the fire. When near the door, he swung the long fringes of his head-dress across the fire, and ran out. The Fire people pursued him. He gave his head-band to Antelope, who ran and passed it on to the next runner. Thus they carried it in relays. The Fire people caught up with the animals, and killed them one by one. (Teit 2)

Coyote escapes and puts fire into the trees and grass, after which the people can use them to make fire.

One might ask what practical information about cheating is presented in this story, given that people have no need to steal fire, and could not use other animals as accessories if they did. Although the plot may be pure fantasy, the social behavior depicted in the story is very true-to-life. For example, Coyote exploits Salishan etiquette in order to advance his scheme (by announcing that the purpose of his visit is to dance, and by insisting that his hosts make the fire bigger). The story also illustrates a use of pretense: Coyote feigns fear when his headdress catches on fire, prompting the Fire people's advice that he back away from the fire, which gives Coyote a pretense for moving toward the door, running in "panic," and making his escape.

The Inuit tale of "Kasiagsak, The Great Liar" (Rink) identifies conditions under which cheating is not acceptable. Kasiagsak is an incompetent hunter who attempts to mask his deficiency with his pathological lying. In one instance, he visits his father-in-law in another village and finds that the household is starving. Kasiagsak boasts that his storehouses are full of food, and offers to bring them provisions from his village. He invites his little brother-in-law to accompany him, but when they get to Kasiagsak's home, there is nothing to eat but one raven. The boy has "to look to the other people for some food; and at his departure the next day, he likewise received all his presents from them, and not from Kasiagsak" (Rink 296). Another time, Kasiagsak hears that a couple in a nearby village has just lost a baby and goes to visit. When he finds out what the dead child's name was, he announces that he has a baby daughter by the same name, which of course is a lie. According to custom, the mourners hold a feast for Kasiagsak and give him gifts to take home to his daughter, which he tells his wife he scavenged from a shipwreck. The next day, the mourners come to Kasiagsak's village to bring him more gifts for his little daughter. When they discover that Kasiagsak does not have any children, they take back all the gifts they have given him. Kasiagsak's end serves as a warning to anyone who would emulate him. One day he finds a piece of whale skin floating on the water, and tells his fellow villagers that he has found a whale carcass. He offers to lead them to it, "but when they had gone a long way from home without seeing anything like a floating whale, they got tired of Kasiagsak, and put a stop to all his fibs by killing him then and there" (Rink 297).

A story may also clarify what constitutes cheating and/or what does not. The Jicarilla Apache tale "Coyote and the Badly Treated Dog," (Opler) is a case in point. In this tale, Coyote helps a neglected dog trick his master into treating him better. Coyote pretends to steal one of the master's chickens, whereupon the dog pretends to chase him. After a bout of pretend fighting, Coyote releases the chicken and pretends to run off. Seeing all this, the master decides that the dog is useful and resolves to take better care of it. In this case, the implied contract between the farmer and dog is, "If you do something useful for me, I will feed you." The farmer hasn't kept his end of the bargain because, as far as he can see, the dog isn't doing him any service; the dog hasn't kept his end of the bargain because he hasn't had any service to perform (or, as he has grown weak from hunger, he hasn't had the energy to do so). This story makes a fine point about reciprocal altruism: you can't blame a person for not reciprocating your favor if there has as yet been no need or opportunity for him/her to do so. Free riding doesn't occur until the need or opportunity arises for reciprocation and the debtor doesn't reciprocate. This is the case with the badly treated dog: by withholding food, the farmer is treating the dog as if he were a free rider, when in reality he simply hasn't had the opportunity (or energy) to reciprocate.

As noted above, cheating involves manipulating one or more of the conditions that must be met to solve the free rider problem, the first of which is that individuals must encounter each other recurrently. One can't reciprocate a favor if one never sees one's benefactor again. And, conversely, one can't be punished for cheating (i.e., excluded from future exchanges) by a person whom one never sees again. Coyote capitalizes on this fact in a Jicarilla Apache tale in which he shows how he can lie (Opler). In this story, Coyote comes upon a camp and agrees to teach the people there how to lie if they give him a horse, blanket, saddle, and whip. After receiving these goods, Coyote simply gallops away on the horse. (Of course, the beauty of the story is that, in tricking the people, Coyote does—technically—show them how to lie, thereby fulfilling and violating the contract at the same time!)

The second condition that must be met for solution of the free rider problem is that individuals must be able to distinguish one conspecific from another. Without such a mechanism, it would be impossible to keep track of who has reciprocated a favor and who has not. One way a free rider can thwart this system is by using disguise or by otherwise pretending to be someone he/she is not. Tellingly, the trickster is often a shape shifter (Hynes). In a Karok story, for example, Coyote is floating down river and sees some girls on shore hooking out driftwood. He transforms himself into a piece of driftwood so that they will pull him out. His plan almost works—two girls in succession hook him but then throw him back into the water, saying "'It might be Coyote'" (Kroeber and Gifford 163). Instead of transformation, the trickster may simply use disguise. For example, the Pawnee tell a story in which Coyote conspires with Fox to get some of Cougar's meat. Coyote tells Fox, "'I have been to the cougar's tipi, and he has much meat in his tipi, and the only way we can get some of it is for me to strap you on a board, call you my son, take you down from my back and tell him that his baby is hungry, that it is his son'" (Dorsey 266). This scheme takes advantage of the obligation a man customarily has to provide food for his wife and children: as Cougar's "wife," Coyote is owed food. Coyote plans to eat the food that is "owed" him and leave, but the scheme backfires on him because, as Cougar's "wife," Coyote is obligated to sleep with Cougar. Cougar is quite satisfied with the (presumably anal) sex, but Coyote is so humiliated that he leaves Cougar's residence as soon as he can, and even attempts to kill Fox to prevent him from telling others about the misadventure. The story thus serves as a warning to potential free riders: if you cheat others, you'll be punished, just as Coyote gets his in the end.

Of course, in small, face-to-face hunter-gatherer societies it would be hard to fool someone through the use of disguise, but there would certainly be opportunities for a person to pretend to be someone else. Band membership is very fluid, with individuals and families frequently leaving one group to visit kin and friends in another. Visiting is a chief means by which foragers cope with resource scarcity: when food or water begins to give out in their territory, they visit family or friends in another camp, in hope that resources will be more abundant there. The sense of obligation between hosts and guests is very strong, and is frequently extended to friends-of-friends. For example, I could show up in your camp claiming to be a maternal cousin of your paternal cousin Vinnie, and you would be obligated to welcome me into your camp, even though no one in your camp—including you—has ever met me. It is easy to see how such a system might be exploited.

A third weapon against free riding is a tally-keeping mechanism, whereby individuals are able to track whom they have done favors for, whom they have accepted favors from, and whether or not those favors have been reciprocated. As with the others, however, this mechanism can be foiled. In a story told by the Pawnee, for example, Coyote is starving and desperately searching for food when he sees seven buffalo approaching. He immediately thinks, "Now, what can I do to these buffalo, so that they will give me something to eat?" (Dorsey 263). He begins singing a song in which he relates how he saved the buffalo from being ambushed:

> You buffalo, you had a narrow escape;
> the people were about to surround you.
> I ran right among them,
> took my knife out, cut their bow-strings,
> went among those who had guns,
> broke the flints so that they could not shoot.
> The people, not being able to shoot returned home.'"
> (Dorsey 263)

Coyote further warns them not to travel east, because there is a big village there. Hearing this, the buffalo think, "'Why, we should have been killed if it had not been for that fellow there'" (Dorsey 263), and they give Coyote meat in gratitude. In reality, there is no village, and there was no ambush attempt. Coyote has managed to get himself reciprocated by manipulating the buffalos' favor-tallying mechanism—that is, by making them think he has done them a great favor when, of course, he hasn't done anything for them at all.

A fourth defense against free riding is punishment: cheaters can be excluded from future exchanges. Of course, it is better never to be cheated at all, and one way to discourage others from trying to trick you is to warn them what will happen to them if they do. Trickster tales are one way of doing this: in many tales, the trickster is punished or otherwise suffers as a result of his/her trickery. The Yanomamö, for example, tell a story about a stingy mother-in-law who cheats her son-in-law out of food. The latter gathers *wapu* fruits and puts them in the river to soak (*wapu* fruits are hard and bitter and must be soaked in water to make them palatable). A few days later, when the man asks his mother-in-law if the fruits are ready to eat, she lies and says they need more soaking. In reality, she wants them all for herself and has been eating them on the sly. When the son-in-law catches her, he throws a sharp rock at her, at which point she turns into an agouti (an animal known by the Yanomamö for its gluttony) and runs off into the jungle. In real life, of course, free riders cannot be transformed into animals, but the point is made: cheaters may be attacked or banished from the group.

The Inuit make this point even more vividly. In "The Blind Man Who Regained His Sight," a mother takes advantage of her son's blindness to cheat him out of the meat he has killed. The youth has his mother aim his bow for him at a bear that is thrusting its head through the window of their hut. He shoots and kills the bear, but his mother tells him that he has missed because she doesn't want to share the meat with him. His sister tells him that he has indeed killed the bear, and slips him meat when their mother isn't looking. One day, the young man asks his sister to guide him to a lake, where he encounters a loon. In marked contrast to the mother's behavior, the son makes and keeps a contract with the animal: "'Give me back my sight, and I will give you my knife in payment'" (Rasmussen, "Netsilik Eskimos"108). His sight restored, the youth decides to go seal hunting, using his mother as a harpoon float. When he harpoons a whale, she is dragged out to sea and drowned.

One might argue that the universality of the trickster theme is a by-product of our evolved interest in such behavior—that the same motivational mechanisms that cause us to pay attention to cheaters and cheating incidentally cause us to share information about cheaters and cheating. However, if humans are designed to exchange a wide variety of information in highly complex ways (and mounting evidence indicates that we are), it follows that at least some of the means by which we store and transmit information are designed as well. Studies of learning in non-human animals suggest that this design includes constraints on information acquisition: specifically, that we are designed to store and transmit some kinds of information and not others (Byrne; Hauser). Cross-cultural themes in oral traditions offer clues to the kinds of information the mind is designed to input socially.

Free-riding stratagems and the consequences of cheating are by no means the only kinds of social information communicated via narrative. Oral traditions contain information relevant to a wide range of selection pressures exerted by group living, including mating, parenting, kin relations, coalitions, and aggression. These themes are well-documented in folklore motif indexes (e.g., Aarne; Thompson, *Motif*). As with trickster tales, stories that address other social situations augment the body of experiences that the listener may reference, thus improving his/her ability to predict and manipulate the behavior of others.

In ancestral environments, where language and storytelling originally developed, there were no books or archives or databases. The only place to store valuable information was in the memories of individuals. The folklore of preliterate, small-scale societies is thus a record of the information our foraging ancestors needed in order to survive—information relevant not only to navigating the social world (Scalise Sugiyama, "On the Origins", but to foraging (Scalise Sugiyama, "Food, Foragers"), predator avoidance (Scalise Sugiyama, "Lions and Tigers"), wayfinding (Scalise Sugiyama, "Humanizing"), and warfare (Scalise Sugiyama, "The Oral Tradition"). As John Bierhorst writes, myths "are inherited from the past" (1). We should take care not to squander our legacy.

Works Cited

Aarne, Antii. *The Types of the Folktale: A Classification and Bibliography.* Trans. Stith Thompson. 2nd revision. Helsinki: Suomalainen Tiedeakatemia, Academia Scientiarum Fennica, 1961.

Alexander, Richard. "Evolution of the Human Psyche." *The Human Revolution: Behavioral and Biological Perspectives on the Origins of Modern Humans.* Ed. P. Mellars and C. Stringer. Princeton: Princeton U P, 1989. 455-513.

Axelrod. R. *The Evolution of Cooperation.* New York: Basic Books, 1984.

Axelrod, R. and W. D. Hamilton. The evolution of cooperation. *Science* 211 (1981): 1390-96.

Baron-Cohen, Simon. *Mindblindness: AnEessay on Autism and Theory of Mind.* Cambridge: MIT Press, 1995.

Baron-Cohen, Simon, Alan Leslie, and U. Firth. "Does the Autistic Child Have a Theory of Mind?" *Cognition* 21 (1985): 37-46.

Barrett, Clark. "Cognitive Development and the Understanding of Animal Behavior." *Origins of the Social Mind.* Ed. B. Ellis and D. Bjorklund. New York: Guilford, 2005. 438-467.

Beckwith, Martha Warren. *Mandan-Hidatsa Myths and Ceremonies. Memoirs of the American Folk-Lore Society.* Vol. 32. New York: American Folk-Lore Society, 1938.

Bierhorst, John. *The Mythology of North America.* New York: William Morrow and Company, 1985.

Byrne, Richard. *The Thinking Ape.* Oxford: Oxford UP, 1995.

---, and Andrew Whiten, eds. *Machiavellian Intelligence: Social Expertise and the Evolution of Intellect in Monkeys, Apes, and Humans.* Oxford: Clarendon, 1988.

Carpenter, M., K. Nagell, and M. Tomasello. *Social Cognition, Joint Attention, and Communicative Competence from 9 to 15 Months of Age. Monographs of the Society for Research in Child Development.* serial no. 255. vol. 63.4 1998.

Cashdan, Elizabeth. "Territoriality Among Human Foragers: Ecological Models and an Application to Four Bushman Groups." *Current Anthropology* 24 (1983): 47-66.

Chagnon, Napoleon. *Yanomamö.* 5th ed. New York: Harcourt, 1997.

Cosmides, Leda. "Deduction or Darwinian Algorithms? An Explanation of the "Elusive" Content Effect on the Wason Selection Task." Diss. Harvard U, 1985.

Cosmides, Leda and John Tooby. "Cognitive Adaptations for Social Exchange." *The Adapted Mind.* Ed. Jerome H. Barkow, Leda Cosmides, and John Tooby. New York: Oxford UP, 1992. 163-228.

---. "Neurocognitive adaptations designed for social exchange." *Handbook of Evolutionary Psychology.* Ed. David Buss. Hoboken, NJ: John Wiley and Sons, 2005. 584-627.

Dawes, R.M., J.M. Orbell, and J.C. Van de Kragt. "Organizing Groups for Collective Action." *American Political Science Review* 80 (1986): 1171-85.

deGroot, A. D. *Thought and Choice in Chess.* New York: Basic Books, 1966.

Dennett, Daniel. *The Intentional Stance.* Cambridge, MA: MIT Press, 1987.

Dorsey, George. *Traditions of the Skidi Pawnee. Memoirs of the American Folklore Society.* Vol. 8. Boston: Houghton, Mifflin, 1904.

Downs, James. *The Two Worlds of the Washo: An Indian Tribe of California and Nevada.* New York: Holt, Rinehart, and Winston, 1966.

Erdoes, Richard and Alfonso Ortiz. *American Indian Trickster Tales*. New York: Viking, 1998.

Fehr, E. and S. Gächter. "Cooperation and Punishment in Public Goods Experiments." *American Economic Review* 90 (2000): 980-94.

Flinn, Mark, David Geary, and Carol Ward. "Ecological Dominance, Social Competition, and Coalitionary Arms Races: Why Humans Evolved Extraordinary Intelligence." *Evolution and Human Behavior* 26 (2005): 10-46.

Fox, Robin. "Sexual Conflicts in the Epics." *Human Nature* 6.2 (1995): 135-44.

Frith, U. *Autism: Explaining the enigma*. Oxford: Basil Blackwell, 1989.

Goodwin, G. *Myths and Tales of the White Mountain Apache. Memoirs of the American Folk-Lore Society*. Vol. 33. New York: J. J. Augustin, 1939.

Hauser, Mark. *Animal Minds: What Animals Really Think*. New York: Henry Holt and Company, 2000.

Humphrey, Nicholas. *Consciousness Regained*. Oxford: Oxford UP, 1983.

---. "The Social Function of Intellect." *Machiavellian Intelligence Social Expertise and the Evolution of Intellect in Monkeys, Apes, and Humans*. Ed. Richard Byrne and Andrew Whiten. Oxford: Oxford UP, 1988. 13-26.

Hynes, William. "Mapping the Characteristics of Mythic Tricksters: A Heuristic Guide." *Mythical Trickster Figures: Contours, Contexts, and Criticisms*. Ed. W. Hynes and W. Dot. Tuscaloosa: U of Alabama P, 1993. 33-45.

Isaac, Glynn. "The Food-Sharing Behavior of Protohuman Hominids." *Scientific American* 238 (1978): 90-108.

Kelly, R. L. *The Foraging Spectrum*. Washington, D.C.: Smithsonian Institution Press, 1995.

Kroeber, A. L. and E. W. Gifford. *Karok Myths*. Ed. Grace Buzaljko. Berkeley: U of California P, 1980.

Kummer, H. and J. Goodall. "Conditions of Innovative Behaviour in Primates." *Philosophical Transactions of the Royal Society of London B308* 1985): 203-14.

Kurzban, R., K. McCabe, V. Smith and B. Wilson. "Incremental Commitment in a Real-Time Public Goods Game." *Personality and Social Psychology Bulletin* 27.12 (2001): 1662-73.

LeBlanc, Steven and Katherine Register. *Constant Battles: Why We Fight*. New York: St. Martin's Press, 2003.

Lee, Richard. *The Dobe !Kung*. New York: Holt, Rinehart and Winston, 1984.

---, and Irven DeVore. *Man the Hunter*. Chicago: Aldine, 1968.

Leslie, Alan. "The Theory of Mind Impairment in Autism: Evidence for a Modular Mechanism of Development?" *Natural Theories of Mind: Evolution, Development and Simulation of Everyday Mindreading*. Ed. Andrew Whiten. Oxford: Blackwell, 1991. 63-78.

---. "ToMM, ToBy, and Agency: Core Architecture and Domain Specificity." *Mapping the Mind: Domain Specificity in Cognition and Culture*. Ed. L. Hirschfeld and S. Gelman. Cambridge: Cambridge UP, 1994. 119-48.

Lowie, Robert H. *Myths and Traditions of the Crow Indians. Anthropological Papers of the American Museum of Natural History*. Vol. 25 Part I. 1918.

Malijkovic, V. "Reasoning in Evolutionarily Important Domains and Schizophrenia: Dissociation between Content-Dependent and Content Independent Reasoning." Unpublished honors thesis, Department of Psychology, Harvard U, 1987.

Norman, Howard. *Trickster and the Fainting Birds*. San Diego: Gulliver Books, Harcourt Brace, 1999.

Opler, Morris. *Myths and Tales of the Jicarilla Apache Indians. Memoirs of the American Folklore Society*. Vol.31. New York: American Folk-Lore Society, 1938.

Ostrom, E., J. Walker, and R. Gardner. "Covenants With and Without a Sword: Self-Governance is Possible." *American Political Science Review* 86 (1992): 404-17.

Pinker, Steven. *How the Mind Works*. New York: W. W. Norton, 1997.

Price, Michael, Leda Cosmides, and John Tooby. "Punitive Sentiment as an Anti-Free Rider Psychological Device." *Evolution and human behavior* 23 (2002): 203-31.

Radin, Paul. *The Trickster: A Study in American Indian Mythology*. Westport, CT: Greenwood Press, 1956.

Rasmussen, Knud. " Observations on the Intellectual Culture of the Caribou Eskimos." *Report of the 5th Thule Expedition 1921-24*. Vol. 7. 2. Copenhagen: Gyldendalske Boghandel, Nordisk Forlag, 1930.

---. "The Netsilik Eskimos: Social Life and Spiritual Culture." *Report of the 5th Thule Expedition 1921-24*. Vol. 8.1-2. Copenhagen: Gyldendalske Boghandel, Nordisk Forlag, 1931.

Rink, Hinrich. *Tales and Traditions of the Eskimo*. 1875. Mineola, NY: Dover Publications, 1997.

Sato, K. "Distribution of the Cost of Maintaining Common Resources." *Journal of Experimental Social Psychology* 23 (1987): 19-31.

Scalise Sugiyama, Michelle. "Folklore and the Adapted Mind." Paper presented at the annual meetings of the Human Behavior and Evolution Society. Binghamton, New York, August, 1993.

---. "On the Origins of Narrative: Storyteller Bias as a Fitness-Enhancing Strategy." *Human Nature* 7.7 (1996): 403-25.

---. "Food, Foragers, and Folklore: The Role of Narrative in Human Subsistence." *Evolution and Human Behavior* 22.4 (2001): 221-40.

---. "Reverse-Engineering Narrative: Evidence of Special Design." *The Literary Animal*. Ed. J. Gottshall and D. S. Wilson. Chicago: Northwestern UP, 2005. 177-96.

---. "Lions and Tigers and Bears: Predators as a Folklore Universal." *Anthropology and Social History: Heuristics in the Study of Literature*. Ed. H. Friedrich, F. Jannidis, U. Klein, K. Mellmann, S. Metzger, and M. Willems. Paderborn: Mentis, 2006. 319-31.

---. "Humanizing the landscape: Storytelling as cognitive mapping." Under revision for *American Anthropologist.*

---. "The Oral Tradition as a Verbal Artifact: Using Folklore to Reconstruct Methods and Tactics Used in Prehistoric Warfare." In preparation for *Warfare in Human Evolution: An Interdisciplinary Perspective.* Ed. Doug Kennett, Michelle Scalise Sugiyama, Lawrence Sugiyama, and Frances White.

Stone, V., L. Cosmides, J. Tooby, N. Kroll, and R. Knight. 2002. "Selective impairment of Reasoning About Social Exchange in a Patient With Bilateral Limbic System Damage." *PNAS* 99.17 (2002): 11531-36.

Street, Brian. "The Trickster Theme: Winnebago and Azande." *Zande Themes: Essasys Presented to Sir Edward Evans Pritchard.* Ed. A. Singer and B. Street. Totowa: Rowman and Littlefield, 1972. 82-104.

Sugiyama, Lawrence and Richard Chacon. "Effects of Illness and Injury on Foraging Among the Yora and Shiwiar: Pathology Risk as Adaptive Problem." *Human Behavior and Adaptation: An Anthropological* Perspective. Ed. L. Cronk, N. Chagnon, and W. Irons. New York: Aldine, 2000. 371-95.

---, John Tooby, and Leda Cosmides. "Cross-Cultural Evidence of Cognitive Adaptations for Social Exchange Among the Shiwiar of Ecuadorian Amazonia." *Proceedings of the National Academy of Sciences* 99.17 (2002): 11537-42.

Teit, James A. *Folk-tales of Salishan and Sahaptin Tribes.* Lancaster, PA: American Folk-lore Society, 1917.

Thompson, Stith. *Tales of the North American Indians.* 1929. Bloomington: Indiana UP, 1966.

---. *Motif Index of Folk Literature.* Vol. 4. Bloomington: Indiana UP, 1957.

Toelken, Barre. "Poetic Retranslation and the 'Pretty Languages' of Yellowmen." *Traditional Literatures of the American Indian.* 2nd ed. Ed. Karl Kroeber. Lincoln: U of Nebraska P, 1997.

Tomasello, Michael. *The cultural origins of human cognition.* Cambridge, MA: Harvard UP, 1999.

Tonkinson, Robert. *The Mardudjara Aborigines: Living the Dream in Australia's Desert.* New York: Holt, Rinehart and Winston, 1978.

Tooby, John, and Leda Cosmides. "The Past Explains the Present: Emotional Adaptations and the Structure of Ancestral Environments. *Ethology and Sociobiology* 11 (1990): 375-424.

---, and Irven DeVore. "The Reconstruction of Hominid Behavioral Evolution Through Strategic Modeling." *The Evolution of Human Behavior: Primate Models.* Ed. W. Kinzey. Albany: SUNY Press, 1987. 183-237.

Turnbull, Colin. *The Mbuti Pygmies: Change and Adaptation.* New York: Holt, Rinehart, and Winston, 1983.

Whiten, Andrew and Richard Byrne. *Machiavellian Intelligence II: Extensions and Evaluations.* Cambridge: Cambridge UP, 1997.

Wiessner, Polly. "Risk, Reciprocity and Social Influences on !Kung San Economics. *Politics and History in Band Societies.* Ed. E. Leacock and R. B. Lee. Cambridge: Cambridge UP, 1982. 61-84

Wilbert, Johannes and Karin Simoneau. *Folk Literature of the Yanomami Indians.* Los Angeles: UCLA Latin American Center Publications, 1992.

Yamagishi, T. "Group Size and the Provision of a Sanctioning System in a Social Dilemma." *Social Dilemmas: Theoretical Issues and Research Findings.* Ed. W. Liebrand, D. Messick, and H. Wilke. 267-287. Oxford: Permagon, 1992.

CHAPTER 4
Nature versus Nurture in *La casa de Bernarda Alba*.

Federico García Lorca's play *La casa de Bernarda Alba* is set in a recognizably Andalusian milieu, while Lorca himself stands as the Andalusian writer *par excellence*. But within this specific geographical and cultural setting, the play embraces universal values, namely the tragic confrontation between the individual's instinctive striving for freedom and societal repression of this desire, as exemplified by Bernarda. It is the story of human nature versus human nurture, though with an interesting twist. In this play, and indeed in much of García Lorca's work (Lima 137), the roles typically filled by men are played by women, and so the society the playwright crafts serves as a counterpoint to that usually seen both in tragedy and in real life. The play enables us to examine the relationships between nature and nurture in human affairs through a kind of reverse-engineering triangulation. By comparing Lorca's uniquely Andalusian social construction to baseline male-role-typical, non-Andalusian societies, we can determine which elements are affected by this change and which are not. Such a determination also enables us to see which of the elements in the social conflict presented in the play are universal constants of human nature and which are constructs of nurture. More precisely, I propose to compare the behavior of the characters in the play with empirical data regarding human behavior in the areas of religion, sibling rivalry, parent-offspring conflict, and male parental investment in order to see how unique Lorca's *Casa* really is.

The play opens with the ringing of church bells as Bernarda and her family attend the funeral services of her late husband, Antonio María Benavides (symmetrically, the play closes with a death, as well). The maids, not invited to the services, are discussing objectively, relatively unmoved, some of the social aspects of the funeral--priests from all the local towns have come, and the church's physical plant is beautifully prepared, though the bells and liturgy are overdone. What does move the servants is the singing voice of the priest as he performs the final "Our Father." La Poncia, one of the maids, compares him to a famous singing priest, Tronchapinos, whom Lorca modeled on an actual Andalusian priest. La Poncia recalls a performance of Tronchapinos in which "Retumbaban las paredes..." [The walls trembled....] and his singing made one feel that it was "...como si un lobo hubiese entrado en la iglesia" [...as if a wolf had entered the church] (122, all translations mine). Later in the play, Bernarda's stallion produces this same reaction in his efforts to get at the mares. His forceful blow to a nearby wall causes la Poncia to scream "¡Por Dios!" [My God!] while Prudencia, a family friend, exclaims "Me ha retemblado dentro del pecho" [My heart is trembling in my breast] (179). Bernarda orders the horse, which Prudencia anthropomorphizes as "Bregando como un hombre" [struggling like a man], freed from his confinement before "nos eche abajo las paredes" [he knocks down the walls] (179). Neither the stallion's sexual instinct nor the women's reactions to the priest's singing and to the horse's struggle are social constructions. Nor did the women, and certainly not the horse, actually learn these behaviors; they are provided by nature, and so give an insight into the underlying human nature from which religion, among other things, springs.

In this regard, La Poncia later tells one daughter to forego sexual relations with her step-sister's fiancé, saying "...no vayas contra la ley de Dios" [don't go against the law of God] (155). Here religion is a response to instinctive sexual desire. Were there no pre-existing desire in human nature, a "law of God" against it would be superfluous. When Angustias shows her engagement ring to Prudencia, this latter woman observes that it contains pearls which, in her time, signified tears, to which Angustias responds, "Pero ya las cosas han cambiado" [But things have changed] (180), implying human nature can be changed. Adela, spokesperson for human nature in the play and Angustias' step-sister and rival for the object of desire, Angustias' fiancé, Pepe El Romano, counters that "Yo creo que no. Las cosas significan siempre lo mismo. Los anillos de pedida deben ser de diamantes" [I don't think so. Things don't change their meaning. It should have been a diamond ring.] (180). By this she implies human nature cannot be changed, that the eternal triangle of sexual jealousy will indeed continue to elicit tears, as it has always done. At this point Bernarda, the voice of social control in the play, proclaims that things turn out as one proposes they do, though Martirio, yet another sister and rival for Pepe, counters, with a touch of foreshadowing, "O como Dios dispone" [Or as God wills them] (180). This controversy figures as another installment of the nature/nurture debate. And finally, toward the end of the same conversation, Prudencia observes that the important thing regarding Angustias' wedding is that everything turn out all right, to which Bernarda replies that there's no reason why they shouldn't, as church bells ring ominously in the background. The church bells here call to mind not only the social control normally associated with the Catholic Church, but also the innate and agonistic instincts of both the girls and Bernarda that give rise to religious control in the first place.

Previously in the play, Lorca has described the tragic history of another family friend, Adelaida, whose father killed his first wife's husband in order to marry her, and then abandoned her for a second woman, whose daughter he married and with her conceived Adelaida, the second woman dying insane. Martirio takes this series of events as an example of how things in life tragically repeat, a question of fate or destiny beyond human control. And later, when Adela asks Bernarda why people repeat an old saying about things being written in the stars when they see a shooting star, the older woman replies that the ancients knew many things that moderns have forgotten. There are many allusions to water as a symbol for life, instinct, nature, and reproduction. For example, Adela uses thirst for water as excuse to go out to Pepe to satisfy another thirst, sexual in this case. La Poncia's equates the ocean with freedom, while Adela's shamanistic grandmother's equates the ocean with sexual freedom. Conversely, all in the play fear the enclosed water of the well/casa will cause the death of the grandmother. This same parallel between trapped water and trapped people is seen in the oppressive town's lack of free flowing water. In the metaphor of control over water and people versus free flowing water and human nature, there is a subtext of a pre-tragic surrender to nature versus the tragic attempt to control it.

Indeed, looking back to the roots of theatrical tragedy, we can see that the genre has evolved from early attempts to control nature. Examples include the Memphite Drama, treating the death and resurrection of Osiris--the god personifying the self-renovation of Nature--and performed each year on the first day of spring in ancient Egypt, and the Abydos Passion Play, which also treated the death and resurrection of Osiris. Both works date from around 2500 B.C. There are also similar Sumerian, Babylonian, Hittite, and Canaanite examples from this same period (Gaster). According to Aristotle (*Poetics)*, Greek tragedy developed out of the dithyrambs or hymns sung, chanted, and danced in honor of Dionysius—the god representing the fertility of Nature--and many other cultures cite similar mythopoetic origins. The power of the dithyrambs recall the power of the priest's singing, and music in general, which is a human universal (Cross; Brown), to evoke emotion and so, the power of the gods, or of God, as in the case of Lorca's play.

Tragedy was originally a religious endeavor, in the sense that religion relies on emotion to induce people do what they otherwise might not. Starving people in India don't eat cattle, not because they rationally understand that these animals are necessary for next year's rice crop, but because the cattle are sacred (Harris 223-24). The Aztecs in Mexico ate human flesh, not because of a rational understanding of the need for animal protein in a diet otherwise practically devoid of it, but because the flesh was made sacred through sacrifice to the gods (Harris 161-65). As exemplified in Prudencia's emotional reaction to the signing priest, religious commitment seems to be innate, since the poverty of supernatural stimulus is nearly complete. There is nothing in our empirical world that should make us believe any of the stories told by any of the religions (which are found in every human culture), including the one we do believe, if we even admit to believing that, our own.

To those who would explain this condition as learned behavior, studies have shown that, cross-culturally, religious beliefs appear very early in childhood development, before explicit religious education begins, even in children from non-religious households (Barret and Richard; Knight et al.; Evans). Children appear to have an innate proclivity to think in religious terms, which increases with adulthood (Bulbulia). In like manner, they have an innate propensity for language (Enard *et al*; Chomsky; Jenkins; Krietman) and for narrative (Bruner; Donald; Hendriks-Jansen). The human predisposition toward religious beliefs seems to be product of the abrupt change in the human brain which occurred some 40,000 years ago, giving rise to symbolic representation. Carved Venus fertility figurines begin to appear from Western Europe to Siberia, symbolic representation in cave painting, including star maps in Altamira, and the earliest burials of human cadavers, accompanied by clear signs of ritual, all date to this period (Gillespie; Formicole et al.; Pettit and Bader).

If we further examine Greek tragedy, we see that it is chiefly concerned with control and conflict over two related aspects of inheritance: bloodlines and material wealth. And indeed, from *Hamlet* and *Lear,* to *Mourning Becomes Electra*, we can see that tragedy still concerns the control of bloodlines and of material goods, two intertwined factors, since resource control is tied to reproductive success (Betzig). In *Bernarda Alba*, as well, the central issues are controlling genetic inheritance through sexual purity, hence the insistence of Bernarda that "¡Mi hija ha muerto virgen!" [My daughter died a virgin] (199), and material inheritance, thus the importance placed upon Angustias' inheritance as her only charm, "¡más vale onza en el arca que ojos negras en la cara" [better a dollar in the bank than beautiful brown eyes in the face] (148).

The emphasis on female sexual purity begs the question of male paternity. The problem for males is that, the more they invest in raising one child, the less they can invest in other children. And everything invested in someone else's child is that much less invested in one's own flesh and blood. So, historically speaking, most human societies have placed restrictions on females to prevent their extra-marital copulation. These range from chastity belts and eunuch guards to foot binding and clitorectomies. It would appear the instinct to invest in one's own runs deep. In many animal species--including lions, langurs and other species of monkeys, gorillas, and ground squirrels (Trivers)--when a new male takes over a group of females, he immediately kills as many infants as possible, thus causing the females to come into estrus and allowing him to produce his own offspring. We see the same effect in Moses' instructions following the conquest of the Midianites: "Now kill every male dependent, and kill every woman who has had intercourse with a man, but spare for yourselves every woman among them who has not had intercourse" (Numbers 31). In turn, when Odysseus finally returns home, he tells Telemachus to execute his slave women who have had relations with Penelope's suitors: "Smite them with your sharp swords and take all the life out of them, and cause them to forget the lust they had underneath the suitors, mingling in secret" (*Odyssey* 22.442-45). And finally, Martin Daly and Margo Wilson have documented the fact that, in humans, stepparents are far more likely to abuse a child than are natural parents. A child is seventy times more likely to be killed by a stepparent than by a genetic parent.

Until very recently, men could not be certain that the child they were raising was theirs, hence the evolution of instinctive male sexual jealousy. Male concern is not unfounded, since extra-pair matings are common in the animal world, even among supposedly monogamous species. Studies on extra-pair copulation in monogamous birds reveal 10-15% of their offspring are sired by extra-pair males (Petrie and Kempanears). According to the *Ethnographic Atlas*, and the *Atlas of World Cultures*, about 83% of human societies are polygynous

(Murdock). DNA studies in various cities in Great Britain have shown the father listed on the birth certificate of a child was not the actual genetic father 10-20% of the time (Ridley). Over a dozen recent studies have shown fertile women at mid-cycle prefer short-term "sexiness" in men other than their primary partner, effectively disdaining long-term traits such as kindness and intelligence. Those women who rated their primary partner as relatively unattractive were more likely to express greater interest in extra-pair men when these same women were at the height of their fertility cycle. In turn, men show greater jealousy during their female partner's mid-cycle, with men rated as unattractive showing the greatest amount of covetousness (Gangestad). This same human propensity to adultery is illustrated within the first few pages of the *Casa de Bernarda Alba*, which is where we learn of the infidelity of Bernarda's own husband with the maid (123).

In addition to genetic legacy, the characters in Lorca's play also recognize the importance of control over material inheritance. In the opening conversation of the *La casa*, la Poncia says to another maid that Bernarda is stuck with five ugly daughters. Whereas facial symmetry is cross-culturally deemed attractive, the contrary holds for asymmetry, (Cunningham; Langlois and Roggman). Only one of the offspring, Angustias, the oldest, has money, and so the marriages chances of the rest remain slim at best. Thus, it is in the best interest of genetic control for Bernarda to privilege Angustias, who controls the material resources necessary to support any future children and Bernarda's future grandchildren. Bernarda, therefore, attempts to control the natural predisposition of human males, Pepe in this case, toward younger women by controlling the sibling rivalry among her own daughters. If all of these girls were available, he would likely choose the younger. Indeed, in each of the 37 cultures studied by David Buss, men preferred younger women as wives. Thus one of the central conflicts in the play is the conflict between Angustia's material resources and Adela's probable higher fertility.

This conflict evidences several mediating factors. First of all, as is well known, prior to 1800 half of all children died in childhood, and parental discrimination among offspring, by sex and birth order, often determined who lived and who died (Boone; Voland), just as it eventually does in *Bernarda Alba.* Older children are usually more likely to survive, so they are almost universally favored. According to Daly and Wilson, infanticide is widely practiced in traditional societies, but none condones the killing of the elder of two children. And finally, cross-culturally, in situations as in Andalusia where both land and economic opportunities are limited, inheritance by primogeniture is generally practiced in order to avoid the subdivision of family lands (Hrdy and Judge).

The tradeoff between material legacy and genetic inheritance is restated by Lorca in the case of Martirio. One Enrique Humanas had been courting her, but he eventually chose another bride who had more money than Martirio. To make matters worse, this rival was, according to Amelia, "¡Y fea como un demonio!" [Ugly as the devil!], (137), to which Martirio replies/rationalizes, "¡Qué les importa a ellos la fealdad! A ellos les importa la tierra, las yuntas, y una perra sumisa que les dé de comer" [What do they [men] care about ugliness! All they care about is the land, the stock, and having a submissive dog that will cook and clean for them] (137). As it turns out, the real reason Enrique gave up on Martirio was interference from Bernarda, who later explains her reasons to la Poncia, namely genetic and material control: "¡Mi sangre no se junta con la de los Humanas mientras yo viva! Su padre fue gañán" [As long as I have a breath of life in me my blood will never mix with that of the Humanas clan! His father was a common laborer] (170).

Additionally, Bernarda refuses to accept outsiders, something reinforced in her take on the sexual exploits of Paca la Roseta, who openly engages in multiple adulterous relationships. The explanation for "la única mujer mala que tenemos en el pueblo" [the only bad woman in town] is that she--and the males who participated with her--"...no es de aquí. Es de muy lejos. Y los que fueron con ella son también hijos de forasteros" [are not of the town, but rather outsiders, all from far away] (132-33). Lest one blame Bernarda's Andalusian environment for her xenophobia, it should be noted in her defense that social and geographic outsiders are instinctively shunned by essentially all human societies, as our species' history of violent competition for reproductive partners, land, wealth, honor, and life itself amply demonstrate (Brown; Sherif; Pinker). Peaceful humans may well have evolved at some point in the history of mankind, but none survived their more pre-emptive brethren.

The fact that Angustias is frail and almost beyond her childbearing years, or as Magdalena puts it "...está vieja, enfermiza...tiene cuarenta¡" [...she's old, sickly....and forty] (139) and "Lo natural sería que te prendiera a ti, Amelia, o a nuestra Adela, que tiene veinte años..." [The most natural thing would be that Pepe married Amelia, or Adela, who's twenty...] (140), offers a bit of a quandary. How can control of genetic and material inheritance be placed in such a leaky boat as that of Angustias? But la Poncia has the answer, echoing Adela's eternal return comment, that things always signify the same thing, and Martirio's that things in life tragically repeat. She tells Adela, "Tu hermana Angustias es una enferma. Esa no resiste el primer parto. Es estrecha de cintura, vieja, y con me conocimiento te digo que se morirá. Entonces Pepe hará lo que hacen todos los viudos de esta tiera: se casará con la más joven, la más hermosa, y esa serás tú" [Your sister Angustias is sickly. She won't live through her first childbirth. She's old and thin, and I guarantee you she'll die. Then he will do what all the men do, marry the youngest and prettiest daughter, and that's you] (155).

In fact, historically speaking, remarriage with a younger sibling of a deceased wife has been the modal pattern in Andalusian society. Even though the human brain is only one third its mature volume at birth, any larger and it simply won't fit through the birth canal, death during childbirth, especially the first childbirth, is a significant human problem, one unknown among other animals (Linden). Indeed, in a cross-cultural study of 296 peoples, only 24 reported the occasional unassisted childbirth (Trevathan). Given this high mortality rate, men in socio-economically stratified monogamous cultures typically remarry with a sibling of the first wife so as to maintain the original dowry intact (Gaulin and Booster).

The preferential material inheritance afforded Angustias thus enhances her reproductive possibilities, while decreasing those of her siblings, and so generates sibling rivalry. Adela and Martirio continue to insist that Pepe is only after Angustias' money, land, and orchards, while Bernarda curses the will of her late husband, which leaves the bulk of the money to Angustias and hardly anything to the rest of the daughters. Even before the wedding, the other sisters begin to rebel. When la Poncia tells them they'll be sewing morning, noon, and night when Angustias has children, Magdalena, who earlier predicted that she herself would never marry, says "Yo no pienso dar una puntada" [I'm not planning on sewing one single stitch], to which Amelia adds "Y mucho menos criar niños ajenos" [And I'm especially not planning on raising somebody else's kids] (158). And later the competition for genetic control through reproductive access, in this case to Pepe el Romano, heats up, as Martirio steals Angustias' photo of him, and Adela eventually becomes his lover, leading to her tragic suicide at the end of the play, even as Bernarda tries in vain to put out the instinctual fires that break out throughout the *casa*.

At the beginning of the third act, in the case of Prudencia, as in that of a woman called la Librada, we see a foreshadowing of Bernarda's reaction to Adela's insubordination. When Bernarda asks Prudencia where her husband has been, Prudencia reports that ever since he fought with his brothers over their inheritance, he refuses to pass through the door of his house. Regarding a second matter, the disobedience of their daughter, Prudencia reports that her husband refuses to pardon the girl. To this Bernarda replies that "Es un verdadero hombre" [he is a real man] (178) for doing so, and adds that "Una hija que desobedece deja de ser hija para convertirse en enemiga" [A daughter who disobeys ceases to be a daughter and becomes instead an enemy] (178). When the daughter of la Librada has a child out of wedlock, which the younger woman subsequently kills, Bernarda's response is that everyone should unite to kill la Librada's daughter before the authorities arrive, that she pay for her affront to decency. In turn, sibling rivalry arises again as Martirio seconds the idea of killing the infanticide, while Adela, possibly pregnant out of wedlock herself, is devastated by the idea.

Even the idea of siblicide is, of course, not unique to Martirio, nor does it arise as a result of Bernarda's parenting skills. Investigations concerning the influence of parenting, based on studies of twins and non-twins who have either grown up together or been separated at birth, suggest that about 40% of one's personality is genetic, 35% comes from non-shared environment, and only 5% is associated with growing up in the same family. The other 20% of the total sampling is associated with measuring error (Loehlin). Sibling rivalry has been documented in mammals, birds, amphibians, fish, insects, and even plants: the victim is almost always the youngest member of the brood (Mock and Parker; Mock). In animals it is instinctive. African black eagles lay two eggs to increase the odds one hatches. Siblicide is obligate, so the first to hatch always pecks the second to death (Mock). In humans the actual killing of a sibling may be rare, but the instinct for violent competion is not. And as noted above, this instinct does not come about as the result of bad parental influence, but is rather the stuff of human nature. As Aristotle noted, it is also the essence of tragedy, think Cain and Abel, Jacob and Esau, Niles and Frazer, and all the other examples seen in tragedy of family relations gone bad. Children are most violent in toddlerhood; it comes to them naturally, and only later are they partially socialized (Holden).

Bernarda's attempts to control her daughters, her efforts to enforce sexual purity and to eliminate sibling rivalry, bring to light issues of parent-child conflict. Amelia refers to Bernarda as a "esa crítica que no nos deja vivir" [critic who won't let them live] (135), la Poncia calls her "¡Mandona! ¡Dominanta!" [a bossy, dominating, tyrant] (119), Magdalena "¡Endemoniada!" [the devil] (198) and so on, while Bernarda herself insists to all "Tú no tienes derecho más que obedecer" [you have no right other than to obey] (173) and beats them "¡Todo lo que quiera!" [All I want!] (166) if they disobey. Which they do when sufficiently motivated by reproductive instincts, as is mirrored even in the grandmother's struggle with Bernarda. Initially, the cause of the conflict appears to be Bernarda's dominating nature, but it is actually the reproductive drive in the sisters that precipitates events. Bernarda then reacts to their efforts to seize control of the genetic and material inheritance. Tragedy not only arises from the conflict between instinct and social control, but also from conflicting instinctual demands. These innate predispositions may take a specific sociocultural expression in a given Andalusian *milieu*, as they do in all cultures, though the ultimate cause of these predispositions is human nature, the predispositions humans inherit from their ancestors, proximate and distant.

To put a finer point on it, in nature parent-offspring conflict is generated by the need of parents to maximize the number of surviving offspring, together with the need of the individual offspring to maximize the number of their own progeny, which means they themselves must first survive to reach parenthood, at which point the roles will reverse. Put simply, each offspring wants to monopolize all the resources, food for example, while the parents must divide the resources among multiple individuals to insure the survival of more than one (Trivers). Save for twenty-first-century humans, organisms universally produce far more offspring than survive. Species that put all their eggs in one basket or procreate just one offspring don't do well over time. In trying to control scarce resources, both the parent and the offspring are selfish, each wanting it their way. Given that offspring cannot directly control parental investment, they tend to employ psychological manipulation. Examples include temper tantrums in chimpanzees (Goodall van Lawick), auto-induced convulsions in pelican chicks (Burke and Brown), and a human baby withholding its smile until it gets its way

(Trivers). Parents also employ manipulation, though categorizing their own selfish actions as being in the child's best interest. Studies show that most parents believe they treated their children fairly, though the majority of grown children claim that the opposite was the case (Dunn and Plomin). It is also in the parents' interest that the offspring share with each other, and in the offspring's interest that they not. So parents encourage altruism and sharing, while discouraging selfishness; offspring almost universally resist this type of socialization (Trivers). In *Bernarda Alba* the parent-child conflict represents a struggle innate in all sexually reproducing species, with both sides trying to maximize their own control over the allocation of resources.

In all the cases discussed above-- religion, sibling rivalry, parent-offspring conflict, and male parental investment--one can find evidence that the conflicts are not purely social. These struggles are not simply instances of various cultural artifacts vying for social supremacy, church or family versus the individual, for example. Humans are not born *tabula rasa*. It would seem, based on the empirical evidence, that even such apparently "pure" social constructs as the Catholic Church, the Andalusian family, the idea of sexual purity in women, and social conventions of material inheritance have their basis in human nature, in the predispositions we all inherit. Although our conflicts seem to be simply human nature versus human nurture, it is important to keep in mind that this latter category arises, ultimately, from the former.

Here it is important to note that these predispositions are indeed expressed in societal mechanisms differing from culture to culture, as a function of the environmental constraints of a given society. Throughout the history of humanity, siblings that competed well came to far outnumber those who did not. We know this because universal sibling rivalry is a documented fact (Mock and Parker 1-11). The puppy who politely waits until his siblings have eaten their fill, rather than competing his little heart out for the best spot on mom or for the scrap on the floor, is less likely to become a grandparent than are his more self-centered and more voracious littermates. And if he is a wild dog, he is guaranteed not to produce a new breed of peace-loving wolves. The same can be said of males who raise other men's children rather than their own, of parents who sacrifice all for one child, and religions with no underlying emotional appeal. The classic example is that of the moths of Birmingham, England. Before the ground they landed on was blackened by industrial pollution, it was light-colored, so similarly-colored moths far outnumbered the dark ones that local birds spotted easily. When industrial pollution darkened the ground, the light-colored moths became more susceptible to predation and the relative numbers of light and dark moths were reversed. The point is, the dark moths didn't become dark in order to survive, but rather they survived simply by being dark at the right environmental moment. Siblings don't compete in order to survive, but rather they survive because they compete, indeed, out-compete. And so the fundamental conflicts of tragedy are in essence the fundamental conflicts of human nature. These conflicts do produce social constructions--religion, family, marriage proscriptions--but they do so for underlying ultimate causes, as the social expression of innate, emotional/instinctive predispositions.

To deny this situation one would have to accept the view that humans are the only animal species whose genetic inheritance, itself the central conflict of tragedy, doesn't actually influence our psychological attributes. To say that our genetic history has no influence on our genetic present, which is to say on us, is a position so out of touch with all the evidence as to be indefensible. The new data coming from a variety of scientific fields are not a challenge to literary analysis but rather, as I hope to have shown here, an asset, enabling literary criticism to replace implicit assumptions about human nature with empirical evidence. Bernarda does not survive because she controls her children; she controls her offspring because people who did so before her survived, and the same can be said of those people emotionally moved by religious practices, those instinctively competing with their siblings, or those reluctant to raise someone else's child instead of their own. Any viable description of the social must take human nature into account, and Lorca does a masterful job.

Works Cited

Barret, J.L., and R.A. Richert. "Anthropomorphism or Preparedness? Exploring Children's God Concepts." *Review of Religious Research* 44 (2003): 300-12.

Betzig, Laura L. *Despotism and Differential Reproduction.* New York: Aldine, 1986

Boone, J.L. "Parental Investment and Elite Family Structure in Preindustrial States: a Case Study of Late Medieval-Early Modern Portuguese Genealogies." *American Anthropologist* 88 (1986): 859-78.

Brown, Donald E. *Human Universals.* New York: McGraw-Hill, 1991

Bruner, J.S. *Acts of Meaning.* Cambridge, MA: Harvard UP, 1990.

Bulbulia, J. "Are There Any Religions?" *Method and Theory in the Study of Religion* 17 (2005): 71-100.

Burke, V.E.M., and L.H. Brown. " Observation on the Breeding of the Pink-Backed Pelican *Pelecanus Rufescens.*" *Ibis* 112 (1970): 499-512.

Buss, David M. "Sex Differences in Human Mate Preferences: Evolutionary Hypothesis Tested in Thirty-Seven Cultures." *Behavioral and Brain Sciences* 12 (1989): 1-149.

Chomsky, Noam. *The Minimalist Program.* Cambridge, MA: MIT P, 1995.

Corbin, John. "Lorca's *Casa.*" *MLR* 95.3 (2000): 712-27.

Cross, Ian. "Music and Cognitive Evolution." *The Oxford Handbook of Evolutionary Psychology.* Robin Dunbar and Louise Barrett, eds. Oxford: Oxford UP, 2007. 649-67.

Cunningham, M.R. " Measuring the Physical in Physical Attractiveness: Quasiexperiments on the Sociobiology of Female Facial Beauty." *Journal of Personality and Social Psychology* 50 (1986): 925-35.

Daly, Martin, and Margo Wilson. *Homicide.* Hawthorne, NY: Aldine de Gruyter, 1988.

Donald, Merlin. *Origins of the Modern Mind.* Cambridge, MA: Harvard UP, 1991.

Dunn, J., R. Plomin. *Separate Lives: Why Siblings are so Different.* New York: Basic Books, 1990.

Enard, Wolfgang, et al. "Molecular Evolution of FOXP2, a Gene Involved in Speech and Language." *Nature* 418 (2002): 869-72.

Evans, E.M. "Cognitive and Contextual Factors in the Emergence of Diverse Belief Systems: Creation versus Evolution." *Cognitive Psychology* 42 (2001): 221-54.

Formicola, V., et al. "The Upper Paleolithic Triple Burial of Dolni Vestonice." *American Journal of Physical Anthropology* 115 (2001): 372-9.

Gangestad, Steven W. "Reproductive Strategies and Tactics." *The Oxford Handbook of Evolutionary Psychology*. Robin Dunbar and Louise Barrett, eds. Oxford: Oxford UP, 2007. 321-32.

García Lorca, Federico. *La casa de Bernarda Alba*. Madrid: Cátedra, 1986.

Gaster, Theodor. *Thespis: Ritual, Myth, and Drama in the Near East*. New York: Doubleday,1961.

Gaulin, Steven, and J.S. Boster. "Dowry as Female Competition.' *American Anthropologist* 92 (1990): 994-1005.

Gillespie, R. "Alternative Time Scales: A Critical Review of Willandra Lakes Dating." *Archaeology in Oceania* 33.3 (1998): 169-82.

Goodall van Lawick, Jane. "The Behaviour of Free-Living Chimps in the Gombe Stream Reserve." *Animal Behaviour Monographs* 1 (1968): 161-311.

Harris, Marvin. *Cannibals and Kings*. New York: Vintage, 1991.

Hendriks-Jansen, H. *Catching Ourselves in the Act: Situated Activity, Interactive Emergence, Evolution, and Human Thought.* Cambridge, MA: MIT P, 1996.

Holden, C. "The Violence of the Lambs." *Science* 289 (2000): 580-81.

Hrdy, Sarah B., and D. Judge. "Darwin and the Puzzle of Primogeniture." *Human Nature* 4 (1992): 1-45.

Jenkins, L. *Biolinguistics: Exploring the Biology of Language*. Cambridge: Cambridge UP, 2000.

Knight, N. et al. "Children's Attribution of Beliefs to Humans and to Gods: Cross Cultural Evidence." *Cognitive Science* 28 (2004): 117.

Kreitman, M. "Methods to Detect Selection in Populations with Applications to the Human." *Annual Review of Genomics and Human Genetics* (2000): 539-59.

Langlois, J.H., and L.A. Roggman. "Attractive Faces are Only Average." *Psychological Science* 1 (1990): 115-21.

Lima, Robert. "Missing in Action: Invisible Males in *La casa de Bernarda Alba*." *Bucknell Review* 45.1 (2001): 136-47.

Linden, Herbert. *The Accidental Mind: How Brain Evolution has Given Us Love, Memory, Dreams, and God.* Cambridge: Harvard UP, 2007.

Loehlin, J.C. *Genes and Environment in Personality Development*. Newbury Park: Sage, 1992.

McDermid, Paul. *Love, Desire, and Identity in the Theatre of Federico García Lorca*. London: Tamesis, 2007.

Mock, Douglas W. *More than Kin and less than Kind: The Evolution of Family Conflict*. Cambridge: Harvard UP, 1004.

---, and Geoffrey A. Parker. *The Evolution of Sibling Rivalry*. Oxford: Oxford UP, 1997.

Murdock, G.P. *Ethnographic Atlas*. Pittsburgh: U of Pitt P, 1967.

---. *Atlas of World Cultures*. Pittsburgh: U of Pitt P, 1981.

Petrie, M., and B. Kempenaers. "Extra-pair Paternity in Birds: Explaining Variation between Species and Populations." *Trends in Ecology and Evolution* 13 (1998): 52-58.

Pettitt, P.B., and N.O. Bader. "Direct AMS Radiocarbon Dates for the Sungir Mid-Upper Paleolithic Burials." *Antiquity* 74 (2000): 269.

Pinker, Steven. *The Blank Slate: The Modern Denial of Human Nature*. New York: Penguin, 2002.

Ridley, Matt. *The Red Queen: Sex and the Evolution of Human Nature*. New York: Penguin, 1993

Sherif, M. *Group Conflict and Cooperation: Their Social Psychology*. London: Routledge, 1966.

Trevathan, Wenda. *Human Birth: An Evolutionary Perspective*. New York: Aldine, 1987.

Trivers, Robert. *Social Evolution*. Menlo Park: Benjamin/Cummings, 1985.

Voland, E. "Differential Reproductive Success Within the Krummhorn Population (Germany 18th and 19th Centuries)." *Behavior, Ecology, and Sociobiology* 26 (1990): 54-72.

CHAPTER 5

Darwin, Sexual Selection, and the Spanish Novel in the Late Nineteenth Century

The widening rift between those who attend to Darwin's language in order to discover reciprocity between his science and its cultural history and those who adapt precepts of evolutionary theory as a means to decipher universals present in all literature, belies a deeper continuity. The shared belief that science and literature belong together makes epistemological posturing counterproductive. When used to complement the humanities and social sciences, Darwin's theories, for all their complexities and unanswered questions, encompass a spectrum of approaches that need not be mutually exclusive. An historical Darwin can also be a universal Darwin, with a balanced say between his own voice and today's evolutionism. It is imperative that he mediate rather than divide our common aim of understanding the reach of representation.

As opposed to the popularized figurehead of the *struggle for life* and an indifferent Nature, my Darwin comes instead from *The Descent of Man, and Selection in Relation to Sex* (1871), where the author takes his observations of the natural world and extends them, though with no shortage of problems, to humankind. Here sexual selection, despite being overshadowed by the polemics surrounding our simian ancestors (disputes which persist even today), stands out to me as Darwin's richest science because it remains his most unresolved. At a pivotal moment in Spanish history, the liberal Revolution of 1868 and the Restoration decades following, womanhood defined by the patriarchal ideal of the domestic angel, *el ángel del hogar*, came under scrutiny as corollary economic and political issues concerning her emancipation began to surface. The gender dynamics of Darwinian sexual selection, and above all female choice, captured the imagination of Spanish realists in a society held by the coetaneous fervor of *el darwinismo*, a nineteenth-century nomenclature born from a need to describe a cultural rather than biological Darwin. In a corrective to evolutionary science as the foundation behind female oppression and the status quo, Darwin and the Spanish novelists whom he inspired illuminate the indeterminacy of the individual and, through the precept of change as a constant, advance the self-determination of women in unprecedented—and often unintended—ways.

Bearing social as well as a scientific import, the theory of sexual selection as expounded in the *Descent* brings with it a singular history of power play between the sexes born from unresolved tensions of female passive choice and selector role reversals. Nevertheless, these elements remain absent from attempts to take a scientific lens to literary representation of gender dynamics in the nineteenth century. It is true, as Alys Weinbaum points out, that the majority of Darwin's contemporaries stigmatized the *Descent* for its association with sex, while in tacit agreement that *On the Origin of Species* (1859) treated the "serious" questions (148). But more needs to be said about the reasons for and consequences of privileging *Origin* at the expense of the *Descent.* Attention to contributions from related fields, evolutionary biology, rhetoric of science, feminism, and cultural studies among them, reveals that the devaluation traces back, in part, to the incongruities of Darwin's conjectures in the latter half of the *Descent.* At the same time, sexual politics play a role, for "sexuality is not only the most intractable element in power relations, but rather one of those endowed with the greatest instrumentality" (Foucault 103). As a catalyst for the scientism of sexuality, Darwinian sexual selection overlaps with corollary feminist critiques of patriarchy, exchange, and otherness. Together, these complementary angles of investigation, straddling nature and culture, delineate a Darwin inseparable from socio-political change.

The quandaries of sexual selection beg an explication, first, of its place in Darwin's larger evolutionary agenda and, second, of its operation. For his part, Darwin considered sexual selection to be a process distinct from natural selection, with success, in the former category, measured by reproduction and, in the latter, by survival. Today, modern Darwinists place sexual selection under the umbrella of natural selection, as one of the two modes (the other being environmental constraints) by which *naturally* achieved variation is maximized. The greater the variation, the greater the chances for variations that will enhance adaptation. The better adapted an individual, the more likely it is that he or she will live to sexual maturity, to mate, and thus to pass on his or her genes to future generations.

However, important distinctions do differentiate natural from sexual selection. When it comes to access to ecological resources, the former may be either inter- or intra-specific, whereas the latter will always occur among members of the same species toward the goal of mating. For this reason, Darwin defines sexual selection as operative according to two dynamics: a competition between two or more members of the same sex (most often males) for the possibility to copulate with one or more members of the opposite sex (most often females); and, a struggle between the sexes where a member of one sex (usually a male) engages directly with a member of the opposite sex (usually a female) in an attempt to reproduce. These two processes, often called "male-male combat" and "female choice," are far less straightforward than the nomenclature implies.

Literature offers an instrumental medium for probing the more troublesome aspects of Darwin's theory. In particular, the late-nineteenth-century novel, because it was so frequently premised on science, was understood by writers of the period to be a science in its own right. In this context, as David Baguley avers, the specific generic features of the novel at this moment in history can be said to derive from science. He maintains that a self-conscious genre program brings renewed uncertainty to the novel, leaving readers "to wonder if the human dramas are there to illustrate the scientific principles or if the science merely serves to motivate the human dramas" (61). Mikhail Bakhtin speaks of this type of crossover in works by Emile Zola, for whom "the activity of coming to know another's word" (353) requires that the "adaptability of the individual" or "testing . . . of the heroes' biological worth" (389) be recognized as the "hybridization" (358) of epic "trial" or "chosenness" (389), of the "*image of a language*" (354), when placed in dialogic interaction with other discourses. These *other* discourses originate in nineteenth-century science, for Bakhtin's characterization of "adaptability" and "biological worth" is, after all, Darwinian.

Along these lines, arguments of "amalgam" and "hybridization," which counterpose objectivity to subjectivity, raise the question of what the "real" of *real*ism actually is. The most useful answers begin with aesthetics as a means to reach the social bedrock of this literary impulse. One way is to think of realism as dependent upon a "silhouetting effect" (Brown 231), which juxtaposes two modes, the dramatic *coup de théâtre* against the atmospheric *tableau*. The consequences of this dynamic are twofold. First, realism emerges as a structure of consciousness, rather than as mere mirroring. Second, individual agency falls under Hegelian terms of contingency in which "reality is a tragic battleground where the active force of the free individual (pure "possibility") always succumbs to the mastering contingencies of historical destiny ('reality')" (236). As to the relationship between self and community, J. Hillis Miller complicates traditional notions of mimetic realism through emphasis on "the *imaginary* quality of reality" and the parallels between the novel as a verbal structure and society as a symbolic game (35-36). Indeed, Brown's structure of consciousness and mastering contingencies together with the generative potential of Miller's *imaginary* quality of reality liberate realism from the empirically real and, so doing, open new space for "real" reciprocity between literature and science.

In late-nineteenth-century Spain, science became synonymous with progress. Thus, José Ortega y Gasset's pithy formula, "Europa = ciencia" [Europe=science} ("Asamblea" 102), from 1908, speaks volumes about the social, political, and ideological implications embedded within this "disinterested" epistemology after the Revolution of 1868 and into the twentieth century. Ortega had no doubts about the need for Europeanization as the method by which to solve "el problema español" (99) [the Spanish problem]; he worried, instead, about the efficacy of the programs in place to achieve such an end. But in a later lecture on Galileo, Ortega better clarifies what he means by science. He proclaims: "La ciencia es, en efecto, interpretación de los hechos. Por sí mismos no nos dan la realidad, al contrario, la ocultan, esto es, nos plantean el problema de la realidad" [Science is, in effect, interpretation of the facts. By themselves they do not give us reality, on the contrary, they hide it, that is, they plant the problem of reality] ("Galileísmo" 15). Ortega posits that science operates according to two distinct processes: "Una puramente imaginativa, creadora, que el hombre pone de su propia y libérrima sustancia; otra confrontadora con lo que nos es el hombre, con lo que le rodea, con los hechos, con los datos. La realidad no es dato, algo dado, regalado – sino que es construcción que el hombre hace con el material dado" [One purely imaginative, creative, which man enacts from his own free constitution; the other in confrontation with that which is for us man, with that which surrounds him, with actions, with facts. Reality is not a fact, something given, bestowed as if a gift – but instead is a construction that man makes with the material given] ("Galileísmo"16). As consequence, truth is never "lo que vemos, sino precisamente lo que no vemos" [what we see, but rather precisely what we do not see] ("Asamblea" 101), like the imperceptible wavelengths of light that paint the world around us.

Nowhere is this situation more pronounced than in the "inasible" [ungraspable] individual, for Ortega asserts that personal character "no sufrirá jamás garantías de exactitud" [will never suffer from guarantees of exactitude] (101-2). Yet if George J. Becker is correct, that realism "denied that there was a reality of essences or forms which was not accessible to ordinary sense perception" (6), Ortega's thoughts must fall far from the realist tree. It is the combination, though, of Ortega's precepts that makes them relevant, for together they define an imaginative science of the unknowable individual beyond the reach of empiricism, and yet still within its epistemological parameters. As a result, Becker understates the scope of the movement. Rather than with *On the Origin of Species* (1859), the Darwinian revolution for realism occurs in 1871, the year Darwin published his first scientific treatise on humankind, moved by what Ernst Cassirer has called our *symbolic imagination* (21-6). With respect to gender relations, the Spanish novel thereafter shows how "abnormal" his discoveries in the *Descent* truly are, as new, disquieting paradigms emerge.

Given the ideological implications of Darwin's theory of sexual selection, one might imagine that in late-nineteenth-century Spain discussions surrounding Darwinism and the "woman question," topics of paramount public interest, would converge. Yet sexual selection remains absent from histories of *el darwinismo* in Spain, and what connection does exist between the theory and coetaneous issues of women's rights, *la cuestión de la mujer*, is not self-evident. Accounts describe Darwin in the context of the liberal Revolution of 1868, a climate characterized by institutional reform, a nascent free press, and urban modernization. Made synonymous with determinism, his theories became a polemical powder keg for both reformists and conservatives. The period remains defined along these lines, with the social consequences (ethical and economic) of natural selection from the *Origin* on center stage during a series of early debates (1868-76), followed by a mostly religious reaction against man's simian ancestry in the wake of the *Descent* in Spanish (1877-88).

Sexual selection may not have brought about women's emancipation in Spain, which is still an ongoing process, but it did contribute to how the "woman question" was discussed within and toward the realization of a civil society. Yet this truth is by no means self-evident. Influential figures in late-nineteenth-century Spain knew Darwin's work; they knew his theories of natural selection, of man's descent, and, as it turns out, of sexual selection. The distinction between the three lies in the fact that whereas the first two topics, natural selection and man's simian ancestry, generated an impressive body of polemics, impressive for its quantity, but not always its quality, the latter does not even appear on the bibliographical map. This silence, though, speaks volumes about sexual selection in Spain. The *Descent* reached the Peninsula, but rather than cause a revolution or act as a one-dimensional touchstone for controversy, it infiltrated the culture without ever being named. This phenomenon, in turn, says as much about sexual selection as a theory, as it does about Spain. Neither entity is clear-cut when it comes to power relations between the genders. The nebulous history of Darwin and women's emancipation in Spain reveals the complex ambiguity and material potential of a theory that was anything but absent from the Spanish imagination.

Jacinto Octavio Picón (1852-1923) provides an ideal point of departure toward an overview of the *Descent* and selected late-nineteenth-century Spanish novelists. His works capture the plasticity of suitors and selectors, the gendered agents of Darwin's model. In particular, the theory of sexual selection reveals the biological underpinnings of the Don Juan figure in the fiction of Picón. At the same time, this author's representation of women, clouded by domestic angels and pyrrhic victories, reflects the unresolved science behind Darwin's principle of female "passive" choice. Yet Picón's writing, noted for its ambivalence with respect to women's emancipation, also shows stability; true conquest in courtship depends on conscience. Moving beyond the carnal to the psychological, Picón understands that the ultimate seduction of Don Juan by his female counterpart is inseparable from Darwinian "moral sense" (71), that is, the ability to reflect on past actions with dissatisfaction and to feel remorse. In the final analysis of Picón's works, economics and appearances give way to a female individualism born from trial and oriented toward what Darwin calls the "highest stage in moral culture" (101).

Though much of the *Descent* treats female selectors as the norm, Darwin cannot reconcile what he sees in nature with observations of his society: "Women are everywhere conscious of the value of their beauty; and when they have the means, they take more delight in decorating themselves with all sorts of ornaments than do men. They borrow the plumes of male birds, with which nature decked this sex in order to charm the females" (372). As a result, though he never says so explicitly, Darwin suggests that in this social environment male (passive) selectors act as the objects of female (active) suitors. The suitor-selector bifurcation, which remains unresolved to the very end, breaks the continuity he seeks to establish between the animal kingdom and mankind. Indeed, sexual selection proves an "extremely complex affair" (296) in which sexual dimorphism—among men and women—allows for transformations that appear anything but natural.

Over the course of his career, Picón would work through Darwin's contradictions in what now appears, from our present vantage point, to be a cohesive progression. In his early novel, *Juan Vulgar* (1885), Picón's protagonist, Juan, lives according to the Don Juan ideal, while parody illustrates the social obstacles that have come to confuse the "exertion of a choice" (Darwin 402) by one sex or the other. Duped into a lackluster marriage by a cunning father, Juan finds himself bored to tears by domestic bliss. His face shows "doloroso asombro" [painful astonishment] (351) when he arrives home one evening and, as if looking in a mirror, is terrified to find his submissive wife, Pilar, *el ángel del hogar*, sound asleep after having attempted to read his life's work. From such reflections on self-alienated male selectors, Picón then probes the power dynamics of courtship in his most acclaimed novel, *Dulce y sabrosa* (1891), which follows the tribulations of Cristeta as she falls victim, in the work's first half, to the consummate seducer, Don Juan de Todellas. In the novel's second half, however, Cristeta devises a secret plan to win him back through, among other strategies, suggestive adornments, and then rejects his marriage proposal: "¡Eso…jamás!" [That…never!] (348), when she succeeds in winning his heart. Indeed, Picón delivers a tale worthy of Darwin's wavering, in which a passive female selector becomes an ornamented active suitor, only to turn in the last instant into a selector anew and on her own terms. Yet it is not until *Juanita Tenorio* (1910) that Picon's exploration of Darwinian inversions comes full circle. In this, his last novel, a hapless first-person narrator named Juanita suffers the abuses of many men in a psychological journey that culminates with a renewed trust in the moral sense of conscience and an uncomfortable conquest of her former, disloyal lover in the interest of Darwinian "family ties" (*Descent* 391).

On a similar trajectory, but with results distinct from those of his lesser known contemporary, Leopoldo Alas "Clarín" (1851-1901) exposes what is most ironic, and hence most ethical, about the rivalries and rituals of sexual selection. If common ground exists within the critical conversation surrounding the literary production of Clarín, particularly of those studies that treat either or both of his two novels, *La Regenta* (1884-5) and *Su único hijo* (1890), it is the recognition that irony foils attempts to find univocal meaning in his authorial voice. In the specific case of analysis of his two principal works, role-playing and gender ambivalence constitute key issues encompassed by irony. What remain untouched, however, are the mechanics that govern the interdependence of these three components of Clarín's fiction and, as a corollary, the function of irony as a constitutive dynamic rather than as solely an agent of destabilization. Through the rivalries and rituals of courtship, Darwinian sexual selection, a dynamic in its own right, provides a formal structure well suited to the interplay between irony, performance, and gender because, in general terms, reproductive impulses infuse Clarín's plots.

Of the two fundamental laws of sexual selection, male-male combat and female choice, the former gives rise to ample discussion in the *Descent* concerning the causes and consequences of rivalry. Speaking of sexual differentiation, Darwin posits that "weapons of offence and the means of defence possessed by the males for fighting with and driving away their rivals" (257-8) were born out of the pressures applied by sexual rather than natural selection. It is not insignificant that at an earlier point in the text Darwin counts language among these developments: "When we treat of sexual selection we shall see that . . . this power [voice] would have been especially exerted during the courtship of the sexes, serving to express various emotions, as love, jealousy, triumph, and serving as a challenge to their [male] rivals" (56), and later, "the vocal organs were primarily used and perfected in relation to the propagation of the species" (330). In effect, these secondary sexual characters "depend on . . . rivalry," for "when we behold two males fighting for the possession of the female . . . and performing the strangest antics . . . we cannot doubt that, though led by instinct, they know what they are about, and consciously exert their mental and bodily powers" (258). A reciprocal relationship materializes between male rivalry and "stronger passions." The resultant modifications define the male as "the more active member in the courtship of the sexes" (272). Rivalry defines the male character: "Man is the rival of other men; he delights in competition, and this leads to ambition which passes too easily into selfishness" (326); and ultimately, "there should be open competition for all men; and the most able should not be prevented by laws or customs from succeeding" (403) in the quest for offspring. In sum, sexual selection produces rivalries between males and, at the same time, would not be effective without the passions, antics, and language that originate and evolve because of contested females.

Exploring the generative desire of rivalries and the virtualities of ritual, Clarín explodes the ethical irony of intersubjectivity, where unfinished individuals are bound together in a community of reinterpretations. He authors sexual selection in the literary realm: his two novels together illuminate what is to be found latent in Darwin's limited awareness of the operative mechanics that fuel the "most complex affair" of courtship. Both works turn on oppositions, which merge and re-emerge out of the psychologies of individuals who craft a material environment driven by the performance of Darwinian "strange antics" and "display." And still, despite the overlooked continuities, in the final scene of *La*

Regenta an indifferent Nature overpowers language, once language has made palpable, through Celedonio's perverse kiss on the lips of the fallen Ana, "el vientre viscoso y frío" [cold and viscous underbelly] (II, 598) of its own destructive potential. Only acerbic silence remains after this testimony to Darwin's caution against the courtship "ambition which passes too easily into selfishness" (326). With *Su único hijo*, though, Clarín intones another tale, this time about reproduction, the unfinished individual, and a language that has rewritten sexual selection by choice. With respect to the child, Bonifacio orchestrates his final lines in the third person, saying to Serafina: "Bonifacio Reyes cree firmemente que Antonio Reyes y Valcárcel es hijo suyo. Es su único hijo. . . . ¡Su único hijo!" [Bonifacio Reyes firmly believes that Antonio Reyes y Valcárcel is his/her/their son. He is his/her/their son] (324). Indeed, Antonio is his/her/their only son, for where thinking is ethical, actual decisions matter less than deciding.

These Darwinian inquiries into the writings of Picón and Clarín remain incomplete without attention to the possible outcomes tied to the variables and mechanics of courtship, namely the heirs and errors of sexual selection. The title of Benito Pérez Galdós' *La desheredada* (1881) gives an immediate sense of the singular place the theme of heredity within the literary production of Spain's most acclaimed novelist (1843-1920). Whether as a tableau for characterization or as a key to the social stratification of nineteenth-century Spain, inheritance enables multifarious readings. As a result, the absence of any scholarship that might seek to bridge this work by Galdós with the *Descent*, where Darwin treats heredity at length, calls into question the unspoken consensus that determinism in Galdosian realism traces back to evolutionary science.

In truth, unresolved issues of agency in Galdós's fiction, particularly with respect to his female characters, mirror the indeterminacy found in Darwin, who appeals to the "scientific" imagination of his public when making the case for his laws of inheritance. Moreover, the theory of sexual selection ties actual deeper social fears to heredity, namely the concern that reproductive errors result from disadvantageous matches as determined by class. Hence, sexual selection unmasks why Galdós, from a cultural standpoint, elects disinheritance as his subject; the plight of Spain, in moral and political disarray during the Restoration, cannot be divorced from the limitations imposed on female choice. Through the collapse of Isidora's identity in *La desheredada*, Darwinian heredity facilitates a reformist ideology inseparable from Galdós' innovations in genre. Yet by the time *Fortunata y Jacinta* was published (1886-87), it is Galdós who uses his subtitle, "dos historias de casadas" [two stories of married women], to show how the plasticity inherent in Darwinian thought begets the regenerative authorship of a nation and its symbolic heir.

With all its mysterious operation and effects, heredity concerned Darwin in no small measure. In the *Descent* his attention to the subject, though sustained, also comes across as disconnected, showing a preoccupation with possible outcomes at the cost of a sound explanation of how personal traits carry over from one generation to the next. In the first half of his tome, Darwin makes inferences about heredity from his personal observations and experience. He opens with a hierarchical model: "we may trace a perfect gradation from the mind of an utter idiot, lower than that of the lowest animal, to the mind of a Newton" (106); then, he speaks about inherited attributes: "with man . . . it is too certain that insanity and deteriorated mental powers . . . run in the same families" (111). Shortly thereafter, he introduces the little understood phenomenon of reversions, "the reappearance of long-lost characters" (113), which can cause arrested development or even deformations (121-27). Darwin also gives an account of the formative role of environment: "There can . . . be no doubt that changed conditions induce [in the individual] an almost indefinite amount of fluctuating variability, by which the whole organisation is rendered in some degree plastic" (114). Change, a constant in the Darwinian model, results from a process of adaptation in which (reproductive) success depends on the use or disuse of certain faculties. It is important to note that this operation *might* effectuate corresponding modifications in the offspring: "Whether . . . modifications would become hereditary, if the same habits of life were followed during many generations, is not known, but is probable" (117). In sum, from disquieting hierarchies to open-ended questions about nature versus nurture, Darwin's initial presentation of heredity teems with ambiguities.

Out of uncertainties about the mechanics of heredity emerge concrete concerns regarding the broader social implications. In terms evocative of Galdós' fascination with the state asylum of Leganés, pathology, and the repercussions of deviance, Darwin worries about degeneration, because civilized nations impede natural selection due to their exaggerated compassion: "With savages, the weak in body or mind are soon eliminated . . . We civilized men, on the other hand, do our utmost to check the process of elimination; we build asylums for the imbecile, the maimed, and the sick; we institute poor-laws; and our medical men exert their utmost skill to save the life of everyone to the last moment" (168). However, failure to fulfill such humanitarian responsibilities would be catastrophic: "If we were intentionally to neglect the weak and helpless, it could only be for a contingent benefit, with a certain and great present evil" (169). More than mere documentation of Darwin's reflections, the slippages of these passages suggest something unnatural, even injurious, about altruism, while the language reveals how these issues of community were discussed in the move toward modernization across Europe. In the shadow of Malthus, Darwin hopes that inferior individuals will not marry, a trend already maintained by Nature (that is, by the economic disadvantages of the lower classes). Most troublesome are the "black sheep" (173), society's perfidious members. No check, though, to their propagation need be instituted, for "in regard to the moral qualities, some elimination of the worst dispositions is always in progress . . . Malefactors are executed, or imprisoned . . . so that they cannot freely transmit their bad qualities. Melancholic and insane persons are confined, or commit suicide. Violent and quarrelsome men often come to a bloody end" (172). Yet "progress is no invariable rule;" for without these balances, "the nation will retrograde" (177). In this way, the trajectory and oversight of heredity extend to the organic well-being of the State, a correlation to which Galdós was very much attuned.

It is true that in *La desheredada* Galdós mocks the claims of his protagonist, Isidora Rufete, to noble parentage; the "facts" prove that she is not the daughter of the deceased Virginia de Aransis. However, as Hegel might say, so much worse for the facts, for at the structural level, the story ends quite differently, in that a lineage between the two can be affirmed. Naturalism's prostitute, who at one time called herself Isidora (now the sentimental shadow of the past), remains the contiguous, metonymic daughter of the marchioness of the *folletín*, because the generic relationship is one of origin and cause. When Isidora visits Miquis out of desperation in the last days before her ultimate fall, she shudders to find a skeleton in his office and asks: "¿De quién serían esos pobres huesos?" [Whose poor bones might those be?] (394). He replies: "Son de mujer. Quizá una tan hermosa como tú... Mírate en ese espejo" [They're of a woman. Maybe one as beautiful as you... Look at yourself in that mirror]; to which she adds: "Tus espejos son muy particulares" [Your mirrors are very peculiar] (394). Hence, the "moraleja" (503) of *La desheredada* resides not in the narrator's final chastisement of those who aspire to fly with "alas postizas" [false wings] (503), where Stephanie Sieburth and others sound a Galdosian allegiance to the status quo, but rather in the romantic, Darwinian conscience of the protagonist. The rightful (maternal) heir to (paternal) disinheritance, Isidora's paradoxical bones, like those of the naturalist novel, are a "very peculiar" mirror indeed. In a twist of literary history, the romantic idealism decried by Galdós has come to support its transgression. This skeleton in the realist's closet, so to speak, reflects the novel's past as no other.

As Darwin makes clear in his discussion of heredity, heirs and errors go hand in hand when it comes to sexual selection. In *Fortunata y Jacinta*, errors in the pursuit of progeny define the plot, as Juanito Santa Cruz ends up with the infertile Jacinta, only to have one child and then another out of wedlock with Fortunata, who in her own right endures a loveless marriage to the impotent Maximiliano Rubín. However, Galdós' narrative success depends on Fortunata, the one character without a family line. The narrator makes clear that if Juanito Santa Cruz were not to have made a visit that day to the bedridden Estupiñá, "esta historia no se habría escrito" [this story would not have been written] (60). Where Juanito ends up is "Madrid primitivo" [primitive Madrid] (61), face to face with Fortunata, her mouth full of raw egg, in the most explicit display of sexual selection in all of Galdós' fiction, a scene in which she:

> ...tenía pañuelo azul claro por la cabeza y un mantón sobre los hombros, y en el momento de ver al Delfín, se infló con él . . . que hizo ese característico arqueo de brazos y alzamiento de hombros con que las madrileñas del pueblo se agasajan dentro del mantón, movimiento que les da cierta semejanza con una gallina que esponja su plumaje y se ahueca para volver luego a su volumen natural.
>
> [had the clear blue scarf over her head and a shawl over her shoulders, and in the moment of seeing the *Delfín* (Juanito), puffed herself up with it . . . so as to make that characteristic arching of arms and rising of the shoulders with which Madrid girls from the *pueblo* offer a welcome from within the shawl, a movement that makes them look like a hen that fluffs up her plumage and gives herself airs only to return then to her natural size.] (62)

Primordial, Fortunata answers the doubts of Darwin. As a Darwinian contradiction incarnate, she offers a *tabula rasa* much like the one upon which the *always already* story of evolution continues to be written. In a telling sentence, Feijoo reveals her plasticity: "Será siempre lo que quieran hacer de ella los que la traten" [She will always be whatever those who get involved with her want to make of her] (659). For Darwin, advantageous reproduction is not Nature's teleology; the paradoxical "end" of Nature is variation, constant change, out of which some forms thrive for being better suited to concurrent conditions. In this light, Galdós takes Juanito from "oliendo a pueblo" [smelling like the people] (10) to procreating with their angel. He breaks down the social stratification of Madrid through sexual selection and, in a political *tour de force*, makes Juanito worthy of his democratic name (13). At the same time, however, Galdós literally impregnates Fortunata with "pensamiento puro" [pure thought], for she carries the novel within her. Readers inherit her language, the "*yiá*" song of "tal ave " [that bird] (63), full of Darwinian errors of understanding.

Moving beyond the operation of sexual selection as it comes to shape and be shaped by the novels of Picón, Clarín, and Galdós in ways Darwin never could have anticipated, it is important to take measure of what this theory from the *Descent* can tell us about broader socio-political concerns of the period. To this end, we turn to Emilia Pardo Bazán (1851-1921). Today, questions about Pardo Bazán's fiction elicit questions about her feminism. The issue is no longer the degree to which her novels conform to her theoretical writings on naturalism, but rather what the ideological agenda of her life and political essays say about the movement for women's emancipation at the end of the nineteenth century. Still, matters of genre remain integral to readings of alternative models for Spanish womanhood in Pardo Bazán, that is, of models that break with the traditional role of domestic angel as assigned to women by Spain's patriarchy. In the 1890s, Pardo Bazán published two novels, *Doña Milagros* (1894) and *Memorias de un solterón* (1896), as complementary pieces in a series entitled *Adán y Eva (Ciclo)*, and I would argue that few, if any, of her creative works capture so well the tensions that went along with being a woman writer intent upon reform in a culture where any *literata* was perceived as a threat to the male-dominated status quo.

The first of these novels, *Doña Milagros*, tells the story of an incompetent father, Benicio Neira, and the demise of his household, which collapses under the weight of his loveless marriage to Ilda, who resents him for his loss of patrimony and, in the end, expires from the exigencies of eighteen pregnancies and multiple offspring. Doña Milagros, the childless wife of a retired army major living in the apartment next door, offers salvation to Benicio through her innate maternal virtue, which he idolizes to the point of gifting her the twins that an aging Ilda bore him just before she died. The sequel, *Memorias de un solterón*, charts the inner awakenings of a self-satisfied bachelor, Mauro Pareja, as his sympathies for the Neira family lead to a genuine love for the bookish, "ugly ducking" daughter of their clan, Feíta (a play on the Spanish diminutive for both faith and ugly). This woman rejects his marriage proposal, only to counter with one of her own in the final scene, a proposal based on greater sexual equality and motivated by a sense of obligation to family, namely, caring for her wayward and penniless sisters. In the context of these plots, feminist critiques identify three principal gender subversions on the part of Pardo Bazán: 1) inversions of sexuality, namely of a "feminine" Benicio and a "masculine" Feíta, 2) usurpation of patriarchal authority through first-person narrations of the respective male protagonists, Benicio and Mauro, 3) and, the emancipatory terms of the marriage contract between Mauro and Feíta.

Though the texts bear out claims about progress toward women's liberation, such assertions also overlook, downplay, or reject the more problematic, or what could be called the "anti-feminist" or conformist readings, as allowed for or even at times encouraged by Pardo Bazán's conclusions. Doña Milagros, a paragon of idealized femininity in the mold of *la mujer virtuosa*, gives closure to the novel as a surrogate mother made heroine, while Feíta, in accepting Mauro, exchanges her aspirations for self-determination for a defeatist sacrifice of self. In both cases, the logic of these approaches has boxed itself in. Either these unresolved aspects of the novels are capitulations to patriarchy (in which case Pardo Bazán is not as progressive in her fiction as we would like her to be) or else they can be reshaped in such a way so as not to appear concessions at all (in which case Pardo Bazán emerges as even more progressive than we could have imagined). With the former situation, the author's personal history and feminist writings on women's emancipation make it difficult, if not impossible, to maintain such ideological disharmony between her novels and coetaneous political essays written in support of civil rights, employment, and education for women. With the latter, we run the risk of letting our own critical agendas skew what the texts permit until we are satisfied that what they do in fact portray is what they *should* portray. Thus, the options are either an unconvincing discord between the author and her works or interpretations that leave us with a sour taste, given the liberties they take.

Paradoxical logic grants the possibility that Pardo Bazán's conciliations, which must be admitted, act as subversions, rather than negate them. In other words, what feels least feminist about these novels need not be so at all. Pardo Bazán's vision for modern Spanish womanhood, which she adumbrates in *La mujer española* (1890) and *Una opinión sobre la mujer* (1892), among other articles, takes the measure of Spain's patriarchy in unique ways. Via a transnational, self-conscious perspective informed by Darwinian deductions whereby natural selection for Spanish females, *struggle for life*, morphs into materialist sexual selection, *trapping a husband*. The issues that emerge surrounding women's emancipation in Spain at the end of the nineteenth century translate to those of more contemporary feminist debates on reproductive labor and domesticity within capitalism, whether seen from a Marxist perspective or not. The fate of Ilda from *Doña Milagros*, determined by Spain's marriage contract and her wage-less labor of no apparent use-value in the procreation of eleven daughters and a deadbeat son, calls into question the ideology of love that colors the closing exchange between the narrator and his maternal soul mate next door.

Corollary discussions of the family structure in the context of Enlightenment notions of the duties assigned to wives for the good of civilization paint Feíta and *her* choice at the end of *Memorias de un solterón* in a light far less beautiful than has been the preference of readers who elect to see her character's transformation in an ironic juxtaposition to her name. In truth, the union of Feíta and Mauro, brought about by ugly necessity and in contradiction of Pardo Bazán's own credo on realism, taints the sentimental beauty of the marriage plot. In these novels, the oxymoronic law of female passive choice in Darwin's theory of sexual selection emancipates only to the extent that it leads us to question the reality of such choice.

Though my time spent with Darwin has made me aware of how widespread is his association with determinism, I suspect that in Spain this misinformed correlation runs deeper than perhaps anywhere else. A disconcerting hunch, no doubt, but at the same time this reputation also reconfirms for me that Spain is the perfect place to look for his redemption. Where the fall is greatest, so too the glory of resurgence. I speak in such tones so as to set the stage for a sort of heroism, for Darwin too long has played the villain in Spanish literary culture. Indeed, this notoriety traces back to his nineteenth-century political defamation, at the moment when his science threatened faith and the ecclesiastical and even the political establishments, but the English naturalist's Spanish history is likewise, literally, of epic consequence. In *Meditaciones del Quijote* (1914), one of José Ortega y Gasset's most influential and enduring works, Darwin lands on the receiving end of a blow that would come to frame the past and future perception of his negative role:

> Las ciencias naturales basadas en el determinismo habían conquistado durante los primeros lustros el campo de la biología. Darwin cree haber conseguido aprisionar lo vital – nuestra última esperanza – dentro de la necesidad física. La vida desciende a no más que materia. La fisiología a mecánica. . . . No hay libertad, originalidad. . . . Adaptación es sumisión y renuncia. Darwin barre los héroes de sobre el haz de la tierra.

> [The natural sciences based on determinism had conquered the field of biology during the early years of the last century. Darwin believes to have managed to imprison all that is life – our last hope – within physical necessity. Life descends to nothing more than matter. Physiology to mechanics . . . There is no liberty (or) originality . . . Adaptation is submission and renunciation. Darwin sweeps heroes from the face of the earth.] (400)

Darwin's science is the beginning of the end of humanity, of all that is noble in our struggle. It is the struggle, though, that Ortega y Gasset overlooks in evolutionism; he writes off Darwin, making him the bane of our existence and creative spirit, without ever understanding his message. For Ortega y Gasset, heroism amounts to the act of willing, "la raíz de lo heroico hállase, pues, en un acto real de voluntad" [the root of the heroic act may be found, then, in a real act of will] (392), and this will, what moves the hero to tragedy for want of lost epic, for want of self, has no place in Darwin's model.

Nothing could be further from the truth. If evolution, which is temporal, leaves us with one thing, it is the (con)science that *being* is *becoming*. Our intuition makes plain, because we will, even toward tragedy, that no self is given. We continue to run up against the "stone wall" (5) of natural laws, just like Dostoevsky's narrator from underground, but we do so by choice, with the intuition that to run in such a way is also natural. The paradox of this Darwinian ontology cannot help but be metaphorical; in the Derridean spaces of *différance* in-between what is and is not, we like our species are defined by differentiation. Likewise, we realize ourselves in language, which originates in sexual selection, while evolution makes clear that we are as much products as producers of our environments. Material change is a matter of will, and *mater* heroism need not be incompatible with the scientific imagination, where symbolic revolutions take place by Nature and out of the indeterminate individual.

Custom told the Spanish people that women should be domestic angels, slaves to their husbands and political non-entities in the larger patriarchal social structure. The conscience of many people found a different sense of identity, one founded on original principles of self-determination. Innate knowledge of ideal beauty rejected the oppression of women, but to find the beautiful, realism took the ugliest route. This contradiction, I think, gets closest to our fascination with the movement; late-nineteenth-century writers describe for us just the sort of reality we should not want to see, while we relish every minute detail. No one was keener on conscience and the double-edged enterprise of realism than Clarín. He looked at his culture and, in its reflection, saw the plight of women emanating from such sources as *La perfecta casada* (1583), Fray Luis de León's canonical conduct book for married women.

In a response of contraries, Clarín gives us two short stories, "La perfecta casada" (*Relatos sueltos*, undated) and "La imperfecta casada" (*Cuentos morales*, 1896). The first tells of a man who commits suicide because his wife is too angelic. The narrator explains the public's reaction after learning of the husband's last words, "Me mato por no aguantar a mi mujer" [I'm killing myself because I can't stand my wife], from a letter he wrote just before death: "El buen sentido hizo que el público en masa, conocidas las cualidades de la virtuosa señora, declarase que aquel hombre se había vuelto loco de pura felicidad doméstica. Sólo así se explicaba el absurdo de *matarse por no aguantar a la perfecta casada*" [The good sense of the public at large, knowing the qualities of the virtuous woman, told them that the man had gone mad from absolute domestic bliss. Only in such a way could the absurdity of *killing oneself for not being able to endure the perfect wife* be explained] (671, italics original). The second, a sketch of the "imperfect" wife, relates the thoughts of a middle-aged woman who, disillusioned with a failed marriage, goes to the theater alone. She comes to see that, having lost all her charm, she is to blame for the way things are. At home again, alone, she rebukes herself for not loving her husband: "Y pensaba la infeliz, mientras velaba esperando al esposo ausente, tal vez en una orgía: '¡Dios mío! ¡Dios mío! La verdadera virtud está tan alta, el cielo tan arriba, que a veces me parecen soñados, ilusorios por lo inasequibles'" [And the unhappy woman thought, while she stayed up waiting for her absent husband, who was perhaps at an orgy: 'My God! My God! True virtue is so divine, Heaven so above me, that at times the two seem to me to be dreams, illusory because unattainable'] (612). Indeed, Darwin and the Spanish novelists require that we look for change where choice cannot be explained, where our intuition tells us reality has gone awry.

Works Cited

Alas, Leopoldo. "La imperfecta casada." *Cuentos morales. Obras completas.* Madrid: Castro, 1995: 2: 607-12.

---. "La perfecta casada." *Relatos sueltos y narraciones incompletas. Obras completas.* Madrid: Castro, 2003: 3: 669-72.

---. *La Regenta.* Ed. Juan Oleza. 2 vols. Madrid: Cátedra, 1994.

---. *Su único hijo.* Ed. Carolyn Richmond. Madrid: Espasa-Calpe, 1979.

Baguley, David. *Naturalist Fiction: The Entropic Vision.* Cambridge: Cambridge UP, 1990.

Bakhtin, Mikhail M. *The Dialogic Imagination: Four Essays.* Ed. Michael Holquist. Trans. Caryl Emerson and Michael Holquist. Austin: U of Texas P, 1981.

Becker, George J., ed. *Documents of Modern Literary Realism.* Princeton: Princeton UP, 1963.

Brown, Marshall. "The Logic of Realism: A Hegelian Approach." *PMLA* 96 (1981): 224-41.

Cassirer, Ernst. *An Essay on Man: An Introduction to a Philosophy of Human Culture.* New Haven: Yale UP, 1944.

Darwin, Charles. *The Descent of Man, and Selection in Relation to Sex.* 1871. Princeton: Princeton UP, 1981.

Dostoevsky, Fyodor. *Notes from Underground.* Trans. Michael R. Katz. New York: Norton, 1989.

Foucault, Michel. *The History of Sexuality.*Vol. 1. *An Introduction.* Trans. Robert Hurley. New York: Vintage, 1990.

Miller, J. Hillis. *The Form of Victorian Fiction: Thackeray, Dickens, Trollope, George Eliot, Meredith, and Hardy.* Notre Dame: U of Notre Dame P, 1968.

Ortega y Gasset, José. "Asamblea para el progreso de las ciencias." 1908. *Obras completas.* Madrid: Revista de Occidente, 1961: 1: 99-104.

---. "Galileísmo de la historia." *En torno a Galileo.* 1933. *Obras completas. Op. Cit.*: 5: 13-20.

---. *Meditaciones del Quijote.* 1914. *Obras completas. Op. Cit.* 1: 309-400.

Pérez Galdós, Benito. *Fortunata y Jacinta*. Madrid: Hernando, 1979.
---. *La desheredada*. Ed. Germán Gullón. Madrid: Cátedra, 2000.
Picón, Jacinto Octavio. *Dulce y sabrosa*. Ed. Gonzalo Sobejano. Madrid: Cátedra, 1990.
---. *Lázaro* [y] *Juan Vulgar*. *Obras completas*. Vol. 6. Madrid: Renacimiento, 1918.
Weinbaum, Alys. *Wayward Reproductions: Genealogies of Race and Nation in Transatlantic Modern Thought*. Durham, NC: Duke UP, 2004.

CHAPTER 6

Consilience of Human Evolution and a *Seemingly* Trivial Detail in the "The Story of the Captive" by Cervantes

Don Quijote is one of the most studied works of literature of all time, as can be easily evidenced by the hundreds of references that appear in any bibliographic search. There are no signs this interest is flagging, and the fact that new theoretical approaches can still find applicability in *Don Quijote* is testimony in and of itself to the genius of Cervantes. In this paper, I will attempt to apply concepts taken from the study of human evolution to a seemingly trivial detail in "The Story of the Captive" in the hope of enriching our understanding as to why *Don Quijote* endures as a masterpiece.

The analysis to follow will borrow heavily from the interdisciplinary discourse articulated in the 1998 book *Consilience: The Unity of Knowledge* by the world-renowned naturalist, Professor E.O. Wilson of Harvard University. Although the word "consilience" may be unfamiliar to most people, the term is hardly a neologism. Wilson traces its origins to the 1840 synthesis *The Philosophy of the Inductive Sciences* by William Whewell, explaining the term as a "...'jumping' together of knowledge by the linking of facts and fact-based theory across disciplines to create a common groundwork of explanation" (Wilson 8). One realizes consilience in a fashion similar to the criterion of reproducibility of results in the experimental sciences. According to Whewell, "The Consilience of Inductions takes place when an Induction, obtained from one class of facts, coincides with an Induction, obtained from another different class. This Consilience is a test of the truth of the Theory in which it occurs" (Wilson 8). Applied to "The Story of the Captive," the test of truth is whether inductions related by Cervantes, three centuries before evolution arose as a scientific theory, can now coincide with inductions generated independently by evolutionary theory. If they can, then there is serendipity in the analysis: *Don Quijote* will endure not just for hundreds of more years but probably for hundreds of thousands of more years---the time span needed for significant human evolution.

Before applying this evolutionary approach to "The Story of the Captive," it would only be prudent to first review a few of its basic theoretical constructs. Fortunately, the fundamentals are elegantly simple even though proof of its mechanics is highly sophisticated. In the early 1980s, physicist Charles Lumsden worked with E.O. Wilson to model mathematically the process by which slight genetic differences could become fixed in a human population due to their adaptability in the culture of that population (Lumsden and Wilson). Inasmuch as genetic differences could accumulate and be selected by the cultural milieu, genes and culture could be said to co-evolve. Like Darwin's original model of natural selection, much of the elegance of gene-culture co-evolution lies in its simplicity. One need not be a biologist in order to understand the process or perceive its logical extension to the arts. Wilson outlines five basic principles:

> During human evolution there was time enough for natural selection to shape the processes of innovation...
>
> The variation was to some degree heritable...
>
> Genetic evolution inevitably ensued...
>
> Universals or near-universals emerged in the evolution of culture...
>
> The arts are innately focused toward certain forms and themes but are otherwise freely constructed.... (Wilson 217-18)

Precisely because the evolutionary model is so simple, and the arts so complex, endless possibilities exist for inferring a plausible adaptive explanation in any given artistic expression. This is the crux of Stephen J. Gould's derision of the approach as "just-so stories." No doubt, many critics will view consilience and "the unity of knowledge" as both a clumsy attempt at academic imperialism and "vaulting ambition" by its advocates (Kitcher 1985). Wilson is conscious of such criticism and takes pains to make clear at the outset what he believes to be the limitations of consilience. The following passage, quoted at some length, reveals both the ambition and the modesty of Wilson's agenda in the arts and their interpretation. Because the passage encapsulates the thrust of Wilson's agenda, it will be used as a template for an evolutionary approach and applied to a seemingly trivial detail, still poorly understood, in "The Story of the Captive."

> The growing evidence of an overall structured and powerful human nature, channeling development of the mind, favors a more traditionalist view of the arts. *The arts are not solely shaped by errant genius out of historical circumstances and idiosyncratic personal experience.* The roots of their inspiration date back in deep history to the genetic origins of the human brain, and are permanent.
>
> While biology has an important part to play in scholarly interpretation, the creative arts themselves can never be locked in by this or any other discipline of science. The reason is that *the exclusive role of the arts is the transmission of the intricate details of human experience by artifice to intensify aesthetic and emotional response.* Works of art communicate feeling directly from mind to mind, with no intent to explain why the impact occurs. In this defining quality, the arts are the antithesis of science.
>
> When addressing human behavior, science is coarse-grained and encompassing, as opposed to the arts, which are fine-grained and interstitial. That is, *science aims to create principles and use them in human biology to define the diagnostic qualities of the species*; the arts use fine details to flesh out and make strikingly clear by implication those same qualities. *Works of art that prove enduring are intensely humanistic. Born in the imagination of individuals, they nevertheless touch upon what was universally endowed by human evolution.* (Wilson 218-19, italics mine)

Inasmuch as Wilson modifies "genius" with "errant," the above passage is especially poignant in the context of both the author Cervantes and his masterpiece *Don Quijote*. Wilson is saying that the "historical circumstances and idiosyncratic personal experience" may indeed be a necessary condition for a masterpiece, just as is genius, but that neither is sufficient. Hence, "solely" is the keyword in the statement "The arts are not solely shaped by errant genius.... " The clear implication is that the artistic work must capture "*universals* or near-universals" that "emerged in the evolution of culture."

A window of opportunity to realize consilience lies in the divergences of the profile of the Captive from the historical evidence of Cervantes own errant life in the military and as a prisoner. One can safely assume that Cervantes diverged from his own "historical circumstances and idiosyncratic personal experience" in order to enhance the quality of the story as measured by the pubic reception to it. In other words, part of Cervantes' genius was to adapt his own experiences to better correspond to a "universal or near-universal" that he correctly intuited. The test of consilience is whether the divergence corresponds to inductions from evolutionary theory or whether the divergence has a more plausible explanation independent of evolutionary theory.

Fortunately, for the purposes of my thesis, a competing analysis already exists that is devoid of any evolutionary theory. In the very opening sentence of the article "Autobiografia y ficción: el relato del capitán cuativo" [Autobiography and Fiction: The Story of the Captive Captain, trans. mine], John J. Allen makes clear the "historical circumstances and idiosyncratic personal experience" of Cervantes in the profile of the Captive. "Reading Cervantes, there are moments when one senses the presence of the author in his work" (trans. mine 149). Allen reduces his thesis to two basic questions: "In what aspects is the life of the author approximated? And why?" (trans. mine, 150). Allen finds only three substantial differences in the profile of the Captive and the historical circumstances of Cervantes:

1) that the captive is a captain and Cervantes was a common soldier who never held rank

2) that the captain falls into the hands of the Turks in Lepanto in 1571 while Cervantes was not captured until 1575.

3) that the date of beginning military service differs; the captive in 1567 and Cervantes, according to his biographers, in 1570 (trans. mine 150).

The first difference Allen quickly dismisses as somewhat uninteresting, its *raison d'etre* being overwhelming obvious: "How does one explain these changes? That of the rank is obvious. The hero of the story, the future husband of the noble and lovely Zoraida, could not be a soldier without rank. It would not sit well with the literary conventions of the times, nor with the style of Cervantes" (trans. mine 150). Then Allen quickly dispatches the second divergence as a necessary adjustment so that the captive could relate the capture of Spaniards in the Turkish victory at Navarino. The rest of the Allen's article meticulously documents various historical accounts that would establish twenty-two years of military service by Cervantes, corresponding exactly to the years given in the Story of the Captive. Allen concludes, quite forcefully, that:

> The story of the captive appears to be the military career of Cervantes, with some small adjustments, easily explained and already cited, and for that reason, the author explicitly clarifies, at the beginning of the fictional account of the captivity, that from there on out, the life of "one so-called soldier de Saavedra" will stand apart. (trans. mine 155)

Assuming Allen is correct in both the inconsequentiality of the date and place of capture and the reconciliation of the years of service, then there is really only one significant difference between the profile of the captive and that of Cervantes: the rank of the soldier. Being the only significant difference gives that difference even greater significance as Cervantes must have thought the value of the alteration far greater than the value of fidelity with his own military career. So, one returns to the first explanation that Allen presents, viz., that a common soldier would not correspond to the literary conventions of the time. This explanation is unsatisfactory to the extent that the rest of *Don Quijote* also does not correspond to the literary conventions of the time. For example, the Cervantist scholar Herschel Brickell writes: "At home, its novelty shocked the conventional-minded, and annoyed the reviewers; it was a literary departure, and it was unmistakably disrespectful to certain honored traditions in literature and society" (v). Even though the elevated rank may have indeed corresponded to literary convention of the time, it seems doubtful that adherence to literary conventions motivated Cervantes to elevate the prisoner's rank from soldier to captain. Another motive must also have existed. The next explanation that Allen offers is likewise unsatisfactory: "it would not integrate...with the style of Cervantes" (trans. mine 150). Such an argument amounts to a tautology: whatever Cervantes writes forms part of "the style of Cervantes." Finally, Allen offers this explanation: " One need only imagine the title: 'The history of the captive soldier.' I don't think so" (trans. mine 150). Ironically, it is in this casual dismissal of the whole question that a window opens for an evolutionary approach.

According to Wilson, "science aims to create principles and use them in human biology to define the diagnostic qualities of the species" (219). What are the diagnostic qualities of the species that can shed light on why Cervantes's changed the military rank of the captive from a soldier to a captain? To answer the question from the perspective of human evolution, one must first recognize that the story of the captive is, in biological terms, also one of mating even if there is no actual mention of any copulation. Mating behavior has long been the domain of evolutionary biology and broad patterns have emerged that correspond almost exactly to what evolutionists would predict based upon the asymmetry in reproductive potential of males and females (the former having almost unlimited sperm while the latter born with a fixed number of eggs). David Buss, a widely cited evolutionary psychologist, explains the genetic economics of mating behavior this way:

A woman who chose poorly---perhaps by selecting a man who would abandon her and fail to help protect and invest in her and her children---would suffer in reproductive currencies compared with the woman who chose wisely (e.g., by selecting a man who would stick around and invest in her and her children). Compare this with the costs and benefits to the less investing sex--in this case, men. Men who made a poor choice and had sex with the wrong women could leave without losing much. They do not bear the burdens of pregnancy... The low level of minimum obligatory investment allows men to fertilize a large number of women, in principle, and so gain a great advantage by pursuing a short-term, relatively indiscriminate mating strategy. One constraint, of course, is that highly discriminating women limit men's pursuit of such a short-term strategy. (410-411)

Extensive survey data very much corroborate the evolutionary calculus of mating behavior and Buss continues under the heading "What Women Desire More.":

> As predicted, women universally desire men with good financial prospects...Across all 37 cultures, women place a greater premium on a man's income than men do on a woman's income. Women also tend to desire characteristics in men that lead to resources over time. What Men Desire More. The study of 37 cultures found only two qualities that men universally desired more than women: youth and physical attractiveness...Youth is a known correlate of reproductive value because women's reproductive value declines sharply with increasing age...there is evidence that men's standards of beauty are highly uniform across cultures...; linked with cues such as smooth skin, clear skin, and a youthful appearance...the premium that men place on physical attractiveness is a proxy for a premium that men place on youth and health, and hence high reproductive value. (421-23)

We can now apply what is scientifically established about human mating behavior to the Story of the Captive. Zoraida first appears not as a person but as a tantalizing jeweled white hand (341) and almost every mention of Zoraida from that moment forward is accompanied by a reference to her beauty---a cue for her high reproductive value. The captive even takes pains to leave no doubt that her beauty is an objective fact and not just the consequence of his own infatuation or sexual deprivation. This should strike a chord with evolutionary psychologists inasmuch as empirical studies now show that beauty is not "in the eye of the beholder" but is "linked with symmetrical features, which signals health and youth" (Buss 423). Zoraida's objective beauty is so extraordinary that it seems to defy credibility. For example, how did Zoraida look when they finally arrived in Spain, after having forever separated from her beloved father, lost her fortune to pirates, hazarded the seas by night in a small boat, and walked along the Spanish coastline barefoot over sharp rocks? Well, the people she meets the next day are "astonished at the Beauty of Zoraida, which at that Instant seem'd to be in its point of Perfection" (363). Not only does everyone *including* Zoraida find Zoraida beautiful, but she is also conscious of the value of beauty itself as evidenced in her first letter to the captive where she identifies herself as handsome, young, and rich and in that order, offering to become his wife once in Spain. Two pages later, Zoraida's assertions are independently confirmed by the renegade: "That evening our renegade came to us, and told us, he had found out that the Master of that House was the same Moor we had been told of, call'd *Agimorato,* extremely rich, and who had one only Daughter to inherit all his Estate. That it was the Report of the whole City, that she was the handsomest Maid in all *Barbary*" (345).

The Captive's sexual attraction to Zoraida is consilient with human evolution and what we know from the evolutionary psychology of human mating. She has the two things that evolutionists suspect a man wants: youth and beauty. The greater challenge for the goal of consilience becomes the symmetrical question: does the Captive have what Zoraida would want as a woman? Recall from the quote above that "women in most cultures place a premium on a man's social status, his ambition and industriousness, and his older age--qualities known to be linked with resource acquisition" (Buss 422). On all three counts the captive fits. He is a *not* a soldier (as was Cervantes in real life) but a captain, he will embark on a risky adventure for his liberty which is a good sign of his industriousness, he is 42 while she, one suspects, is not yet 20.

Perhaps the best evidence for consilience of this seemingly trivial detail of elevated rank is the attitude of Zoraida's father throughout the story. Because the father perceives low status to the captive, he does not entertain the notion that Zoraida could become attracted to the captive. It is a delicious irony that precisely because the captive is a captive, he becomes trustworthy as a man in the eyes of the father: "...the Moorish and Turkish Women never suffer themselves to be seen by any of their own Nation, but by their Husband, or by his or their Father's Command; but as for the Christian Slaves, they let them see them and that more familiarly than perhaps would be wish'd" (Cervantes 349). This is exactly what evolutionists would predict and Cervantes must have induced this irony as he repeats virtually the same theme in the very next page "Moorish Women make no Difficulty of shewing themsleves to the Christian Slaves, she drew near, without Scruple, to the Place where her Father and I were talking; neither did her father shew any Dislike of her coming, but call'd to her to come nearer" (350). What the father does not yet realize is that Zoraida has already disassociated herself from both her religion and her country. Hence her pool of potential sexual mates does not constitute the "Bassas and ViceRoys" of Barbary but the Christian slaves of which the Captive stands out as one of high social status.

The seemingly trivial detail of Cervantes elevating the rank of the captive from soldier to a captain is not trivial at all. Indeed, the asymmetry in perception of the captive's status by father and daughter is key to how the whole story unfolds. The father readily accepts his daughter's feigned fainting in the Garden when she and the slave are caught in a quasi-embrace (352). Even on the boat, when the tables have turned and the father is now the captive, he cannot perceive his daughter's participation in the flight despite the fact that she hides her face in the Captive's hands, an obvious show of shame (356). The father is unsuspecting and pleads with his captors for his daughter's release not knowing the irony of his plea. Offering whatever they wish for her ransom proves emotionally too much for Zoraida, at which time she breaks down and throws herself into her father's arms. Even the slaves, no doubt men hardened by decades of imprisonment, were shaken "...they began again so passionate and tender a Scene, that most of us could not forbear accompanying their Grief with our Tears" (357). By creating sympathy for the very man who held the captives prisoner, Cervantes achieves here what Wilson perceives as the "the exclusive role of the arts", viz., "the transmission of the intricate details of human experience by artifice to intensify aesthetic and emotional response" (218).

The process by which the father realizes that his daughter has deceived him has everything to do with his inability to perceive early on the captive's social status in the eyes of his daughter. On the boat, he puzzles as to how Zoraida could have been captured wearing such fine clothes. It is only after having eyed the small trunk of jewels that he realizes there is no other plausible explanation but that of her complicity. Confirmation comes brusquely from the renegade who tells the father not only has he been betrayed but that his daughter is a convert (357). Quite correctly, the father does not blindly accept the word of a Renegade but confronts the girl, who implores the holy Lela Marien as the source of all answers at which point, the father throws himself into the sea. He is dragged back on board by the slaves, half drowned, and slowly regains consciousness. The crew decides to leave him on a distant shore where he cannot easily alarm his countrymen. From the shoreline, the father hurls insults to the Christians that, by their nature, cohere with his new understanding of the relationship of the Captive to his daughter: "Why do you think Christians, that this wicked Woman desires I should be set at Liberty? Do you think it is because she is not able to bear my presence, which hinders the Prosecution of her ill Desires: I would have you think neither that she has embraced your Religion, because she has heard that she may live more loosely in your Country than at Home" (358-59). According to the scientific findings of evolutionary psychology, such an insult would indeed hurt the captive inasmuch as men "find promiscuity terribly repugnant in a long-term mate" (Buss 424). So, the insult may not really be directed toward his daughter but as a ploy to have her rejected by the captive. This inference is supported by the fact that the father now begs her to return which implies that she is not hostage but acting of her own will and so he forgives everything and begs her to return. She invokes Lela Marien to protect him and continues on her journey.

Allen is correct in that the story of the captive would not work if the captive were just a simple soldier. The drastic lengths to which Zoraida goes "touch upon what was universally endowed by human evolution" (Wilson 219) only if the captive were of an elevated social status, like a captain, and not the simple soldier that was Cervantes in life. A consilience emerges of this seemingly trivial detail--raising the captive's rank from soldier to captain in what is otherwise a faithful autobiographic profile of Cervantes' own military career--and human evolution. Given the painstaking analysis that just this one detail requires, one can only imagine that a consilience of the entire *Don Quijote* and human evolution is an endeavor that will engage scholars for generations.

Works Cited

Allen, John J. «Autobiografía y ficción: el relato del Capitán Cautivo (*Don Quijote* I, 39-41)». *Anales Cervantinos* 15 (1976): 149-155.

Brickell, Herschel. "Introduction to *Don Quijote* by Miguel de Cervantes." de Cervantes, Miguel, *The Ingenious Gentleman Don Quijote de la Mancha*. Ozell's Revision of the Translation of Peter Motteux. New York: The Modern Library, 1930.

Buss, David M. "The Psychology of Human Mate Selection: Exploring the Complexity of the Strategic Repertoire.. *Handbook of Evolutionary*

Psychology, Ed. Charles Crawford and Dennis L. Krebs. Mahwah, NJ: Lawrence Erlbaum Associates, 1998. 405-29.

de Cervantes, Miguel. *The Ingenious Gentleman Don Quijote de la Mancha.* Ozell's Revision of the Translation of Peter Motteux. New York: The Modern Library, 1930.

Gould, Stephen. "Sociobiology and human nature: a post Panglossian vision." *Sociobiology Examined.* Ed. Ashley Montagu. New York: Oxford UP, 1980. 183-90.

Kitcher, Philip. *Vaulting Ambition: Sociobiology and the Quest for Human Nature.* Cambridge, MA: Harvard UP, 1985.

Lumsden, C.J. and E.O. Wilson. *Genes, Mind, and Culture.* Cambridge, MA: Harvard UP, 1981.

Wilson, Edward O. *Consilience: The Unity of Knowledge.* New York: Alfred A. Knopf, 1998.

Chapter 7

Toward a Natural History of the Soul: Cervantes, *Don Quijote* pt. 2, chs. 13-14, and the Culture of *Limpieza de Sangre*[1]

It is a truism of Cervantes scholarship that the author of *Don Quijote* was judicious about revealing—with anything resembling total transparency or complete candor—heterodox opinions concerning the Church or anything even vaguely ecclesiastical. Given the baleful vigilance of the Inquisition, the slightest hint of deviance would constitute defiance and could easily lead to disaster. Untoward relaxation on an author's part might result in relaxation, as execution after condemnation by the Holy Office was called. Nonetheless, many readers have slipped into the intentional fallacy, construing Cervantes' mind into that of his narrators and characters. The novelist's delicate and deft irony, not to mention his devious perspectivism, seem to have kept him out of the Inquisition's nets, at least during his lifetime.[2] After all, censors are not widely known for their perspicacity, though the Inquisitorial mind—as opposed to a genuinely inquisitive one—can certainly read whatever it's looking for into a given text. Which is not to say that Cervantes expresses no ideas that might run contrary to the Holy Office's expressed program. Indeed, his could be a dangerous game, given a potentially lethal mix of talent and testosterone, as complete "innocence" evinces the profoundest irony and vice versa. Following this same vein, it is my contention (along side of many preeminent Cervantes scholars) that don Miguel, not to mention don Quijote, the novelist's chivalric protagonist and sometime *alter ego*, were probably of mixed lineage: that is, Cervantes and the initially unnamed *hidalgo* who by force of will would become don Quijote de la Mancha, are of *converso* ancestry. To clarify terminology, the *conversos* or *marranos* were converts or descendants of converts from Judaism or Islam to Roman Catholicism.[3]

Starting in 1449 in Toledo, those of such lineage were officially condemned to a potentially lethal limbo in which they could hold no governmental or ecclesiastical office, could not attend university, or even emigrate to the New World (once that venue was "discovered"). For our purposes here, the most immediately relevant consequence of the *Estatutos de Limpieza de Sangre* [Statutes of Purity of Blood] was that Christian baptism was *de facto* invalidated: ethnic (or genetic) provenance effectively trumped Providence. Jews and Muslims might be persuaded—by fair means or foul—to receive baptism. Nonetheless, from the very highest levels of the Church hierarchy, the canonical decrees held that this rite could never expunge the stain of Semitic origins. Through all generations of time and even if the original conversion were at a remove of several generations, a person of that descent, *ex illis* as the Latinism indicated, together with his or her posterity, would always be Jewish or Moorish (Castro, *Casticismos* 42-45; Shepard 55-56). A native-born Spaniard, baptized as an infant according to Catholic credence, who could not trace his or her Old Christian lineage at least back to Adam, in essence became an alien in his or her own land, not to mention in his or her own religion. I don't intend to discourse at any length as to how such dogma violates both the letter and the spirit of the teachings of Jesus of Nazareth and the early Christians. St. Paul, himself a Jewish convert to the new faith, who is often called the "Apostle to the Gentiles," states categorically that Christians were "no longer strangers and sojourners, but...fellow citizens with the saints and members of the household of God" (Ephesians 2:19 RSV). But such utopian circumstances would not obtain in Iberia or her territories. At this point, I emphasize how such existential contradictions equate to the fixity of forms forever, cementing a lack of possible evolutionary change. Some two and one-quarter centuries after the publication of the *Quijote*, Darwinism would enter an Iberian world where essentially the same stifling atmosphere still obtained: a miasma meandering through a maelstrom (Sicroff; Márquez Villanueva, *España* 50-58; Alcalá 523-44; Yovel 75-76 *et passim*).

Those Sephardim who refused this dubious baptism were expelled from Spain in 1492 and from Portugal in 1497. Jews who did accept baptism became New Christians forever more; against them and their progeny for centuries to come would be directed the brunt of the Inquisition's ferocity. The Holy Office would not finally and completely be dissolved until 1834, though elements of the Inquisitorial mindset retained potency for many decades thereafter. In turn, the descendants of the Iberian Moors, with redoubled irony often referred to as "the other *conversos*," were finally (re)conquered in January of 1492. Known thereafter as the *moriscos*, these people would finally be expelled from their native land beginning in 1609 and culminating in 1614, hardly a year before the second part of *Don Quijote* went to press. The novel takes place in a nation set upon itself, people against people, with government and Church against all. In general accord with regnant humoral medical theory, Spain was a saturnine society, an autophageous empire bent on consuming its children and then itself from within. But in the very first lines of *Don Quijote*, Cervantes rejects this noxious milieu, where one's genealogy is everything, refusing initially to name the *hidalgo* in question or to specify the village from which he hails (pt. 1, ch. 1). In this regard, the neophyte knight errant will (re)make himself; he is not subject to paternalistic paradigms, as inherited from or imposed by ecclesiastical and governmental authorities. Rather, the newly-minted Don Quijote creates his being according to the literary models he finds in the *libros de caballerías* [chivalric romances], which he reads so assiduously as to go mad. Crazy like a fox, at least in Cervantes' instance, I think! There is, by the way, a cogent contemporary scientific explanation for his lunacy. Cervantes describes his protagonist as a melancholic whose humors are disastrously out of balance, leading to his lamentably comical behavior.[4] The ironic import here is that whatever fantastic patterns he (or we) might encounter in *libros de caballerías* or even in medical tomes on humoral imbalance are frequently no less—and no more!—insane than the ones in effect throughout Iberian domains.

A brief incident that Sancho recounts in chapter 13 of the 1615 *Quijote* may illustrate Cervantes' attitude toward the Inquisitorial regimen under which Spain labored. The anecdote in question may also figure as a parable of contemporary Iberia, where ancestry is everything, not only the pedigree of wines, but also of the people who drink them. In this mix, vintage bleeds over into lineage (Varela-Portas de Orduña 79-100). Don Quijote's squire has been drinking with a recently-arrived counterpart who serves the Caballero del Bosque [Knight of the Forest]. But beneath shifting layers of identity and disguise, the new arrivals' reality is not yet revealed. In the novel, as in the culture at large, any unguarded openness, any casual or unmitigated frankness, could prove lethal. Throughout this chapter and into the succeeding ones, the dramatic irony is palpable, for Sancho, not to mention his master, remains pristinely unaware that his drinking companion is actually his neighbor Tomé Cecial, while the knight who's come on the scene is really Sansón Carrasco. This subterfuge may be Cervantes' variation on the traditional theme of *in vino veritas*, though the truth of the matter does not emerge for some time, and even then remains ambiguous. In recognition of this deliberative, albeit dilatory, mode, I propose paring back additional layers of "innocent" obfuscation to recognize authorial malice—charity?—aforethought.

Throughout the entire novel, Sancho is seldom, if ever, reticent about mentioning his Old Christian heritage. He asserts that he is fit to serve as Governor of an island, given his clearly *castizo* background that is never far from the foreground. Américo Castro describes how the best way to guarantee one's nobility was not to have noble blood, but rather, to be an unlettered and preferably illiterate *labrador*, a field laborer.[5] This, despite a palpable horror of manual labor, generalized throughout all strata of society, as getting one's hands dirty could detract from *hidalguía* [nobility]. In point of fact, the aristocracy, particularly the *hidalgo* class, the lowest rung on the hierarchy of aristocracy, and even royalty itself, were saturated in *converso* blood (Castro, *Realidad* 266-71; Mariscal 48-49; Yovel 67-70 *et passim*). So it is as an "expert" witness that Sancho discourses so fluidly on one lineage and then another, boasting that he enjoys "un instinto tan grande y tan natural en esto de conocer vinos, que, en dándome a oler cualquiera, acierto la patria, el linaje, el sabor, y la dura y las vueltas que ha de dar, con todas las circunstancias al vino atañaderas" (pt. 2, ch. 13; 733) [a good and natural instinct in matters of knowing wines, because just by the bouquet I can tell you the origin, type, flavor, vintage, and how it was decanted, with all of the circumstances of the wine] (500). He has come by this talent naturally, that is, by inheritance: "tuve en mi linaje por parte de mi padre los dos más excelentes mojones que en luengos años conoció la Mancha" (pt. 2, ch. 13; 733) [I had on my father's side of the family two of the best winetasters La Mancha has had in a long time] (500). The squire cites as proof--I use the word advisedly--a trial to which these two winebibbers were put:

> Diéronles a los dos a probar del vino de una cuba, pidiéndoles su parecer del estado, cualidad, bondad o malicia del vino. El uno lo probó con la punta de la lengua; el otro no hizo más de llegarlo a las narices. El primero dijo que aquel vino sabía a hierro; el segundo que más sabía a cordobán. El dueño dijo que la cuba estaba limpia y que el tal vino no tenía adobo alguno, por donde hubiese tomado sabor de hierro ni de cordobán. Con todo eso, los dos mojones se afirmaron en lo que habían dicho. Anduvo el tiempo, vendióse el vino, y al limpiar de la cuba hallaron en ella una llave pequeña, pendiente de una correa de corbobán. Porque vea vuestra merced si quien viene desta ralea podrá dar su parecer en semejantes causas (pt. 2, ch. 13; 733).
>
> [They gave both of them some wine to taste from a barrel, asking their opinion of the condition, characteristics, and goodness or badness of the wine. One of them put a drop of it on the tip of his tongue, and the other did no more

> than smell it. The first one said that it tasted of iron and the second said that it tasted more of Cordovan leather. The master said that the barrel was clean of iron or of leather. Even so, the two famous winetasters maintained what they had said. As time went by, the wine was sold, and when the barrel was cleaned they found a small key attached to a strap of leather inside. And that'll show your grace if someone from that lineage can give his opinion about such things] (500).

This anecdote savors of folklore and popular culture. Most traditions tell tales of drinking prowess—typically involving copious consumption, rather than delicate gustatory competition—from the ancient Greeks (e.g. Plato's *Symposium*) to the more modern, fraternal sort set on or near college campuses. I'm convinced that actual genealogy also figures as a key to understanding Sancho's story's significance, a meaning which has generally been passed over by readers of the *Quijote*. One who did make something of the brief narrative was the Scotsman, David Hume, who in 1757 discussed it at length in his essay "Of the Standard of Taste." He elaborates on the "great resemblance between mental and bodily taste…that beauty and deformity…are not qualities in objects, but belong entirely to the sentiment." Extrapolating on this latter point, we might assert that in like manner Cervantes recognizes that virtue or worth are not inherent in an individual, but is, rather, earned. Hume goes on about the fineness of "organs" and the "delicacy of taste," and how these may be cultivated to the point of "perfection." Again, there's nothing innate in this process, which must be perfected to achieve perfection. The Scotsman further argues that "prejudice is destructive of sound judgment, and perverts all operations of the intellectual faculties," counsel the Inquisitors would have done well to heed. Though Hume has his own biases, contending that "the Roman catholic religion …inspire[s] a violent hatred of every other worship," and that "religious principles are…a blemish in any polite conversation, when they rise up to superstition." Finally, he avers the universality of "principles of taste," that since they are "uniform in human nature," all men are equal or at least akin in their (potential) perceptions. Groups or individuals may differ among themselves even, though there remain significant commonalities. For Cervantes, as for Hume, this understanding constitutes a key to humanity.[6]

As a prelude to my own interpretation of Sancho's vignette, I note in passing that Cervantes sketches out much the same story in one of his *entremeses*, comedic interludes to be performed between acts of a longer dramatic production. Early on in *La elección de los alcaldes de Daganzo* [*The Election of the Magistrates of Daganzo*] (1615), the rural *regidor* [municipal official], Algarroba (a kind of bean), refers to one of the candidates for town council, Berrocal (his name means *hardscrabble*), as a superb judge of wines:

> En mi casa probó los días pasados
> Una tinaja, y dijo que sabía
> El claro vino a palo, a cuero y hiero:
> Acabó la tinaja su camino,
> Y hallóse el el asiento della un palo
> Pequeño, y dél pendía una correa
> De cordobán y una pequeña llave (67-68).
>
> [Why, just the other day he tried some wine
> I had at home. He told me that it tasted
> Of wood and iron and leather. When the jar
> Was drained, imagine what it still contained:
> A piece of wood, from which there hung a strap
> Of kidskin, and on that a tiny key!] (45)

His companions praise such “rara habilidad” [rare ability] and “raro ingenio” [genius], asserting that “Bien puede gobernar, el que tal sabe” (68) [someone so gifted is cut out to govern] (45). Now, before we laugh these country bumpkins and their antics off our stage, it should be recalled that millions of votes were cast, once and then again, for the individual who became and remained the forty-third President of the United States, purportedly because he seemed a lot like a “regular guy,” one with whom other regular folks could sit down and “have a beer.” More recently still, the candidacy of the commonplace for governance by the average—I want almost desperately to say “mean” or even “mediocre”—was reaffirmed. A candidate for Vice President of the United States appealed with considerable, though happily, not ultimate, success to a mythical “Joe Six-Pack” and his beer-drinking buddies who were not so imaginary. In this light, Cervantes, a satirist of such rustic mores, emerges as a prescient pundit, whether in our thoroughly modern age, or in his own.

Having vented my irony for the moment at least, I'll state frankly that the meaning I intend to extract from the stories put in Sancho's and Algarroba's mouths, lies drowned as deeply—though probably not by accident—as the key on the leather strap. I believe that Cervantes deliberately, though ever-so-delicately, imbues a spiritual sense into this discourse on spirits, proffering a parable of the effects of the Statutes of Purity of Blood. Along similar lines, a particular posture in Cervantes scholarship periodically comes to the fore: it is a chauvinistic stance unfortunately never fully obviated, even by Castro's *El pensamiento de Cervantes*, to the effect that the novelist was somehow an “accidental genius,” that he was heedless or otherwise uncomprehending of the myriad of cultural traditions that infused his literature (see also: Márquez Villanueva, *Letra viva* 48-73). Cervantes was nothing if not clever, as well as deliberate in his work. Certainly he was not conscious of every one of the diverse elements that flow into and through his fiction. But in the case of Sancho's story, I'm convinced that he is quite cognizant of the role that the little key on the strap could play. An equal measure of delicacy, together with a dollop of deviousness and a dash of deviltry, are probably indicated for a reader who, like the winetasters in question, would discern the actual import of the story.

Anecdote is elevated to "historia verdadera" [true history], however natural or unnatural this notion may seem in a work of fiction.[7] Though he probably would not have called it such, the novelist depicts in this vignette, as in the story line surrounding it, a natural history of social policy, not to mention of the spirit or the soul, however one defines those entities. Sancho's tale within a tale may not seem on par with the other, more ostensibly substantial intercalated narratives in *Don Quijote*: the *Curioso impertinente* [Tale of Impertinent Curiosity] (pt. 1, chs. 33-35); the story of Marcela and Grisóstomo (pt. 1, chs. 12-14); the history of the Captive Captain and Zoraida (pt. 1, chs. 37-42), though it nonetheless delineates an organic symbiosis of the cultural and spiritual even before there was biology *per se*.[8] Additionally, though Cervantes wrote more than two centuries before Darwin was born, he shows how notions of evolution, of change and development, whether random or teleological, inform humankind. According to the Holy Office, man, albeit most favored of the Creator, was one more inviolate species among all the others. Paradoxically, this perspective also seems to confirm the human kinship with other animals that Darwin would later affirm, whereas in Cervantes' world, groups, ethnicities, races, and identities were officially fixed, *noble* in the chemical sense, decreed unchanged and unchangeable *a priori* and *a posteriori* for all time and into eternity. This stance contradicts Ortega y Gasset's understanding both of Cervantism and of Darwinism. Meditating on the *Quijote* (1914), he casts the English naturalist as more than a bit of the Inquisitor:

> Las ciencias naturales basadas en el determinismo habían conquistado durante los primeros lustros el campo de la biología. Darwin cree haber conseguido aprisionar lo vital...dentro de la necesidad física. La vida desciende a no más que materia. La fisiología a mecánica. El

> organismo, que parecía una unidad independiente, capaz de obrar por sí mismo, es inserto en el medio físico, como una figure en un tapiz. Ya no es él quien se mueve, sino el medio en él. Nuestras acciones no pasan de reacciones. No hay libertad, originalidad. Vivir es adaptarse: adaptarse es dejar que el contorno material penetre en nosotros, nos desaloje de nosotros mismos. Adaptación es sumisión y renuncia. Darwin barre los héroes de sobre el haz de la tierra (400; see also: Close 180; Varela-Portas de Orduña 81)
>
> [The natural sciences based on determinism conquered the field of biology during the first decades of the nineteenth century. Darwin believed he had succeeded in imprisoning life...within physical necessity. Life is reduced to mere matter, physiology to mechanics. The human organism, which seemed an independent unit, capable of acting by itself, is placed in its physical environment like a figure in a tapestry. It is no longer the organism which moves but the environment which is moving through it. Our actions are no more than reactions. There is no freedom, no originality. To live is to adapt oneself; to adapt oneself is to allow the material environment to penetrate into us, to drive us out of ourselves. Adaptation is submission and renunciation. Darwin sweeps heroes off the face of the earth] (164).

One way or the other, Cervantes successfully adapted his discourse (and his life as discourse) to his milieu, assuring his text's survival into Darwin's time and beyond.

Flowing along with the imagery of wine in Sancho's story is that of blood. Such a convergence clarifies early on in the novel with Don Quijote's dubious battle with the wineskins: the spilled wine, for him and even for Sancho for a time, is the blood he's shed of giants (pt. 1, chs. 35-36; see also: Varela-Portas de Orduña 79-100). It is in the Scripture, however, that these liquids become most explicitly and universally coefficients for each other. Jesus offends many of His followers with his literalism, stating that only if they drink His blood, can achieve eternal beatitude with Him (John 6:54). This reference is usually interpreted to augur the institution of the Eucharist at the Last Supper, which, by the way, most writers accept as a traditional Passover celebration. At that gathering Jesus addresses his Apostles—all of whom, by the way, were New Christians—with regard to the cup of wine they will share, calling it "my blood of the covenant, which is poured out for many for the forgiveness of sins" (Matthew 26:27-28; Mark 14:23-25; Luke 22:20). Ever since, the Eucharistic wine has commemorated the blood of Deity, shed as a sacrifice of the One to be eternally renewed on behalf of all who profess Christianity. Moreover, in Roman Catholicism the equivalency of blood and wine is taken to the very letter, as the priest performs transubstantiation, the concrete and unequivocal indwelling of the Divine, whereby the emblem becomes literally the blood of Christ. Along this same imagistic trajectory in the very last book of the New Testament "the wrath of God" is described as "great wine press" from which blood flows copiously (Revelation 14:19-20). The Inquisition would picture itself as treading out "the grapes of wrath," though without spilling the blood of heretics, which was contrary to their priestly vows; therefore lapsed sinners were strangled and/or burned (Yovel 169). In this same matter-of-fact vein, I propose that Sancho's story of wine stands as a parable of blood, wherein even the slightest trace of "impurity" colors—contaminates?—the entire vessel.

Some of Sancho's words, spoken prior to his telling of the story, lead me to suspect ulterior motives, not so much on his part, but for sure on that of Cervantes. The squire's counterpart and drinking companion offers him a *bota* of wine. Sancho gladly accepts and having drunk deeply, sighs and exclaims: "¡Oh hideputa, bellaco, y cómo es católico!" [Oh, the son-of-a-bitch, how good it is!] Apparently not realizing the multiple meanings of such expression, the Squire of the Forest queries: "¿Veis ahí … como habéis alabado este vino llamándole 'hideputa'?" [You see … how you've praised that wine by calling it a son of a bitch.] Their conversation carries a strong aftertaste of irony, paralleling that of the iron key in the vat. Sancho dives into the verbal mix-up, countering: "Digo … que confieso que conozco que no es deshonra llamar hijo de puta a nadie cuando cae debajo del entendimiento de alabarle." [I confess … that I realize that it's not an insult call anyone a son of a bitch when it's intended to praise him.] He then exhorts his host to tell him if, as he thinks, the wine is from Ciudad Real. The other squire answers in the affirmative, praising Sancho's discernment and affirming that this wine "tiene algunos años de ancianidad" (pt. 2, ch. 13; 722) [it has been aging a few years]. Despite Sancho's best intentions, not to mention those of his creator, there arise other possible senses to this vocabulary that may shed light on the narrative in question.

The labeling of the delectable vintage as *hideputa* takes on additional significance when we recall an incident in another of Cervantes' *entremeses*, *El retablo de las maravillas* [*The Marvelous Puppet Show*]. In short, two con men, Chanfalla and Chirinos, come to a small rural town intending to hoodwink the local officials and the townsfolk. The tricksters offer to perform for their hosts a marvelous panorama, one that isn't really there. Chanfalla and Chirinos tell the pretentious and therefore doubly gullible bumpkins that only those of Old Christian lineage and legitimate paternity will be able to see the display. The municipal leaders and other residents take great pains to prove that they're neither crypto-Jews nor bastards and see all sorts of wonders that never were. They make a spectacle of themselves over an imaginary spectacle. It becomes a case not so much of fear of illegitimacy, as it is of the discovery by one's neighbors of that shortcoming. Additionally, the fear of illegitimacy as to one's corporeal parentage, of whether one's father was actually married to one's mother, reinforces and recapitulates the parallel concern over one's spiritual legitimacy as a Christian. Morbid concern over matters beyond one's control, such as birth in wedlock or out, or whatever other aspect of one's ancestry, would consume the nation's vitality and spirit, driving individuals and peoples to extraordinary lengths of concealment. This toxic blend of self-hatred and self-aggrandizement, a self-immolation beyond even Inquisitorial purview, is parodied in Cervantes' comic interlude, which finally isn't so funny. In turn, Sancho, foul-mouthed but possessed of a hyper sensitive palate, says he means no *deshonra*, that he actually intends his epithet as praise. Américo Castro has write at length on the *conversos*' vain (I mean the adjective to indicate both *proud* and *futile*) quest for *honra*. Castro explains that, though the distinction doesn't always hold true, at the time of *Don Quijote*, honor is typically inherent in one's character, something inviolate, unless one decides otherwise. On the other hand, *honra* frequently appears as a function of popular perception, as perceived in the eyes of others. The quest for *honra* becomes Sisyphean, futile and unrealizable by its very nature. By the same token, no one can be dishonored unless he or she allows it. Beneath Sancho's words Cervantes signals his fellow citizens' folly, as they are engaged in a quixotic quest against phantom enchanters of their own device.[9]

Sancho also terms the delicious wine *bellaco*—Lathrop does not translate this word directly—another ironic opposite and perhaps the equivalent of more or less endearing usage such as *little dickens* or *little rascal*. In contemporary English, we encounter such substitution, where, for instance, *bad* becomes *good*. The wine Sancho samples is so good it's bad, recalling perhaps Chanfalla and Chirinos, whose picaresque artistry is so endearing and enlightening to the reader/spectator, who is charmed and conned every bit as much as are the townspeople in the *entremés*. Here we might recall another picaresque work, the anonymous *Lazarillo de Tormes* (1554), where the eponymous narrator's mother's program is "arrimarse a los buenos, por ser uno dellos" (15) [to mingle with the good folk, so as to be one of them] (my translation), a paradigm which her son will follow religiously. In this situation the good people are those with this world's goods. Incidentally, Américo Castro links these categories with an Hebrew expression, the *benei tovim*, *children of the good ones*, who are such not necessarily in a moral sense, but more in regards to their worldly station (*Realidad* 222-23). In turn, when Sancho calls the wine *católico*, meaning *good*, we might be led to see Cervantes' irony behind that of his character, a further doubling of the *double entente*. In Spain, after all, who could guarantee his or her own catholicity, his or her own goodness, which for high- and low-born was always in doubt?

The reference to the vintage's *ancianidad* also has bearing on this configuration. The Squire of the Forest's meaning remains ambiguous, as he may mean *longevity* in the Christian religion as measured by pedigree, in other words, that one is an Old Christian. He could also suggest—albeit at a slant—that *ancianidad*, whether for wine or for wine drinkers, can accrue with faithful performance and the passage of time. Along this same vein, even the word, *confieso*, that Sancho uses when he acknowledges his grasp of what *son of a bitch* means in context, may call to mind the New Christians' predicament. Another of the terms applied to these people was *confes(s)o*, meaning *one who confessed* or *I confess*, which is what the repentant Jews would do of Christ at baptism and the Judaizers cry out under torture by the Inquisition. This same sort of ironic usage appears in the first section of the *Lazarillo*, where the narrator describes his father's confession, not to mention his own along the narrative way (14-15; see: Nepaulsingh 42-48). Nor will Cervantes leave well enough alone. On a fluke, don Quijote unhorses the Knight of the Mirrors (previously the Knight of the Forest, soon to be revealed as Sansón Carrasco), demanding that his prostrate and momentarily unconscious adversary "confesar y creer" [confess and believe] that Dulcinea is the fairest of them all and that the Don Quijote he'd supposedly defeated was not the one holding a sword to his throat now. Coming to his senses in more ways that one, the defeated rival states his "confieso" (pt. 2, ch. 14; 745) [I confess] (508; see also: Unamuno 165; El Saffar 96-98).

Still current today is the debate across the pages of the *Quijote* with regard to the relative virtue of one's birth and one's deeds. In the official culture, there was no way on earth (or in heaven), no matter what one might do, to make up for a less-than-advantageous birth/upbringing. But the country *hidalgo* who will become Don Quijote obviates his origins in favor of great deeds and the resounding new name he chooses (pt. 1, ch. 1). In turn, the wine Sancho praises is good by nature, despite or because of its origins in Ciudad Real, and therefore deserves the label of *católico*. Performance is the key. The squire recognizes, first the goodness as action, what the wine "does," then its place of origin. In this configuration, Cervantes guardedly asserts that not only the wine, but also the drinker thereof, stand as a function of behavior, of becoming, rather than original and innate being. Could it be, as well, that he proposes, however cryptically, that one can be a Catholic if he or she behaves as a Catholic? Incidentally, even the mention of Ciudad Real, a city founded originally as Villa Real by Alfonso X in La Mancha (New Castile), seems to confirm this emphasis on practice, that one is what one does, not necessarily what one was born. For services rendered to the crown during a civil war, Juan II of Castile conferred the distinction "Muy noble, muy leal" on Ciudad Real. This designation was confirmed in 1475, by Juan's daughter, Isabel, and her husband, Fernando de Aragón, the Catholic Monarchs. In this regard, deeds were what finally mattered, not humble origins, which could be altered, whether exalted or denigrated, legitimized or delegitimized, as circumstances and exigencies, popular or even royal, might demand.

After imbibing the goodly wine, Sancho and his as-yet unknown comrade at the *bota* fall asleep. The next morning brings its own disarray, as Don Quijote's squire for the first time gets a good look, up close and personal, at his drinking companion. He is shocked, nay horrified, to see the man's enormous nose. The narrator describes this appendage as follows:

> ... la primera que se ofreció a los ojos de Sancho Panza fue la nariz del escudero del Bosque, que era tan grande, que casi le hacía sombra a todo el cuerpo. Cuéntase, en efecto, que era de demasiada grandeza, corva en la mitad y toda llena de verrugas, de color amoratado, como de berenjena; bajábale dos dedos más abajo de la boca; cuya grandeza, color, verrugas y encorvamiento así le afeaban

> el rostro, que en viéndole Sancho comenzó a herir de pie y de mano como niño con alferencía (pt. 2, ch. 14; 740).
>
> [… the first thing Sancho Panza saw was the nose of the Squire of the Forest, which was so big that it almost cast a shadow over his whole body. It's said, in effect, that it was enormously large, hooked in the middle, and all covered with purple warts, like an eggplant. It extended the width of two fingers beneath his mouth, and when Sancho saw it, his feet and hands began to tremble like a child with epilepsy] (505).

Such hyperbolic description of the "new" squire, along with Sancho's manic reaction to such a "vestiglo" [monster], in the first instance, may parody the cultural ideal of the *lindo cristiano*, one who had no shadow of Semitism in his or her background. Focusing by extension on the grotesque opposite of such an impossible ideal of beauty, Cervantes ironizes both extremes, demonstrating via the absurdity of one, the ridiculousness of the other. *Cuaresma* [Lent] and *Carnaval* [Carnival], poles that evolved together across the Middle Ages and especially in Juan Ruiz' comedic masterpiece, *Libro de Buen Amor* (1330 or 1343), as Arturo Marasso notes (34; see also: Redondo 191-203 *et passim*). Cecial's horrid nose signals the horror of the bigotry that both shuns and begets it. Opposites, though initially repulsive, ultimately attract.

In keeping with the ubiquitous assertion of the *Quijote* as "historia verdadera," material that was equally "true," though patently fictitious would appear in natural histories, in medical tomes, and in acknowledged works of fiction. A case in point are the chivalric romances that Don Quijote believes as gospel truth, from which would issue a phantasmagoria of creatures. Even the scientific treatises held that out in the wide world, not to mention at home, there were—or would be—many diverse species or strange variants of humankind: people with dog's heads, people with no heads, people with tails, and so forth (Huet; del Río Parra). Earlier on in the *Quijote* (pt. 1, chs. 29 to 37), the figure of the anthropoid, typically, an ape or the monkey, draws erect, recalling considerable medieval and renaissance lore (Janson), harbinger of the fears Darwin engendered when he linked mankind to simians in *The Origin of Species* (1859) and later in *The Descent of Man* (1871). The beautiful Dorotea, searching in disguise for vindication after allowing herself to be seduced by Don Fernando, takes on the role of Princess Micomicona, who begs the knight errant to champion her cause against a wicked giant (see: Varela-Portas de Orduña 94-98). By the way, a *mico* is a *monkey*, which then doubled into *cómico*, to come to Micomicón, her native country (Lathrop trans. 227). Mimicry apes mimesis, and vice versa. The overtly belletristic and the profoundly realistic commingle, informing each other and Cervantes' transcription of the Book of Nature where they all reside. Together, they augur the nineteenth century and the poetic realism that is Darwinism (Hyman).

In Cervantes' time and into more recent centuries, such normally monstrous appearances extended to the *conversos*, who were often labeled as animals of one sort or another, including *marranos [swine]*, *perros [dogs]*, and what to me is particularly interesting, *alboraiques* (or *alboraicos*), after Al Burak. This hybrid beast, with traits of the mule, the wolf, the lion, and the human being, among others, in a single night carried Mohammed to Jerusalem and back, as recounted in Sura 17 of the *Qur'an*, and then in the *Libro del Alboraique*, an anonymous anti-*converso* tract from the late 1480s (Gitlitz, "Hybrid Conversos" 1-17; Shepard 24-28; Fuchs 1-20). Following Inquisitorial teratology, the New Christians were neither fish nor fowl, they were neither Jews nor Christians, but were *hybrids*, *mutants*, as we might say today (Gitlitz, "Hybrid Conversos" 1-17; Fuchs 1-20; Márquez Villanueva, *España* 203-27). But, as Cervantes and so many other *conversos* take pains to point out, in terms of cultural production, they were by no means sterile, unlike most hybrids, mules, for example. Moreover, the *conversos* were renowned, at least popularly, for their fecundity. In another of Cervantes' works, the *Coloquio de los perros* [The Dogs' Colloquy] (1613), Berganza, one of two canine interlocutors, compares the procreative proclivities of the "morisca canalla" [Moorish rabble] to that of the sons of Jacob, who went down into Egypt as only a few souls and returned as 600,000 fighting men, without a count of women and children (317-18). This sort of neo-nativist imagery runs together (as it runs down) the two *converso* groups who by their numbers threaten to overwhelm Old Christian folk (Larsen).

The enormity of the proboscis in play reinforces an anti-Semitic stereotype in play even today, that Jews are identifiable by their big noses.[10] Likewise, the mention of the eggplant is by no means innocuous: in the popular mind, this vegetable was a staple food of Jews and Moors, such that a preference for it would constitute another ethnic indicator (Gitlitz and Davidson 48-49). On the one hand, the Inquisition and the State maintained a monolithic mythology, emphasizing a fixity of ethnicities (if not of actual species); that is, once a Semite, always a Semite, and never to grow or evolve into a Goth and an Old Christian. Stasis was the watchword of generations for which there would be no regeneration, ever. *Conversos* and other marginalized groups might be variants on the human model, though shading toward the subhuman, a modality that the Nazis would later perfect. In much the same taxonomic venue, Don Quijote himself is also rather nonplussed by the squire's appearance: "le juzgó por algún monstruo o por hombre nuevo y de aquellos que no se usan en el mundo" (pt. 2, ch. 14; 742) [he thought he was some monster, some new kind of man not from this world], in other words, some new sort or species. Though Cervantes seems to deflate this protoevolutionary mentality, reasserting fixity, at least in the sense of literary genres and authorial progeny: in the prologue to the novel he writes, "no he podido yo contravenir al orden de la naturaleza, que en ella cada cosa engendra su semejante" (pt. 1, prologue; 9) [I haven't been able to violate the laws of nature, which state that each one begets his like] (3). Cervantes seems to follow suit to the Inquisition, one monstrosity that would beget so many offspring in its own likeness, though once more the novelist ironizes all he surveys, insinuating variance, however grotesque and anti-natural, at every juncture.

The novelist and his protagonist continue in a lengthy tradition of Jewish and *converso* comics, from Alfonso Alvarez de Villasandino to Juan Alfonso de Baena to Antón de Montoro to Francesillo de Zúñiga to Francisco López de Villalobos in the fifteenth and sixteenth centuries, to a welter of funnymen and women of the twentieth and twenty-first centuries. From the painful pratfalls of Don Quijote to those of the Three Stooges, from the sometimes not so passively aggressive antics of Lenny Bruce to Don Rickles to Robert Klein, from Woody Allen to Billy Crystal to Adam Sandler, from Gilda Radner to Jack Black to Sarah Silverman, "Jewish humor" is typically self-deprecating, exaggerating rather than minimizing misfortune for comedic effect. Such mirth directs away the full vigor of persecution and prejudice by receiving it full in the face. The comedians and comediennes seem to accept their lot with laughter, even with self-ridicule, though often the one being mocked becomes the mocker. They caricature the Other in themselves and themselves in the Other. One way or the other, the minority becomes a majority, as laughter—with mordant or more gentle (gentile?) irony—is turned back on the audience.[11] A case in point in Early Modern Spain may be the frequency of animal surnames, from Gato [cat] to Truchas [trout] to Zorrilla [fox], from Cabra [goat] to Gallo to Vaca [cow]—perhaps as a response to the vicious epithets?—that the New Christians took as they left Judaism for Catholicism (Shepard 84-86). The Old Christians might attempt to dehumanize their fellow citizens, who, internalized the affronts and returned them, deconstructed, disabled, and defused by apparent or actual acquiescence. So the huge nose that so agitates Sancho not only trumpets Jewishness, it also recalls the prodigious olfactory sensibilities ascribed to the *dogs of God* (a play on Dominicans, the order of many of the first Inquisitors, as *domini canes*) sniffing out Semitic heresies (Reston). The nose in question also mocks Sancho's oenophilic pretensions, his nose for vintage, be it of wines or of human beings.

The crowning irony to the interaction between the two squires comes when Sancho sees his counterpart without the honker, erasing in an instant "aquella fealdad primera" [that which caused his ugliness] that earlier had so dominated their discourse. The now even newer squire pulls out of his pocket "unas narices de pasta y barniz, de máscara, de la manifatura que quedan delineadas" [a nose made of pasteboard and varnish in the shape that has been described], while Sancho finally takes a good look at "el ya desnarigado escudero" [the un-nosed squire]: "¡Santa María, y valme! ¿Este no es Tomé Cecial, mi vecino y mi compadre?" (pt. 2, ch. 14; 744-745) [Holy Mary help me! If it isn't Tomé Cecial, my neighbor and pal] (508). There's a joyful reunion of friends and neighbors, as Sancho realizes that the monster of his imagination was just like him. At the same moment, the triumphant pair, Don Quijote and his squire also discover that the downed knight is another friend and neighbor. What Walt Kelly, the creator of the comic strip *Pogo*, said, "We have met the enemy and they is us" was as true for Early Modern Spain as it is for us today in the early twenty-first Century. Spain may constitute a special, though not necessarily a unique, case, in that the Other is the native son (or daughter). Spaniards must come to terms with the fact that they are the Other and the Other is them (Yovel). The putative enemy is not just among us, he is us, as we are our own worst enemy, who is one of us. I am put in mind of Baudelaire's famous dictum, "mon semblable, mon frère," addressed to his "hypocrite lecteur." Cervantes could easily have meant likewise for his readers with pretensions to *casticismo* [ethnic/racial purity]. The fratricidal contention they perpetuate (and perpetrate) harkens back to the conflicts between Jacob and Esau, the not-so-fraternal twin brothers who became, respectively, the patriarch of Israel and the type of Rome and Roman Christianity (Yuval 10-11 *et passim*).

This exchange takes on added (addled?) significance a few chapters later when Sancho encounters another old friend. Don Quijote's squire is returning from his ill-starred "island"; the Old Christian has not necessarily been as successful at governing as his Old Christian lineage might initially have led him to anticipate. Sancho eats and drinks with a group of pilgrims, one of whom makes known his real identity: he is another friend and former neighbor of Sancho's, Ricote the *morisco,* one of those expelled from their native Spain not long before. Paradoxically, Ricote seems to justify the King's "gallarda resolución" [bold resolution], as he calls it with pungent irony. The *peregrino en su patría* [pilgrim in his own country] has returned from forced exile under penalty of death, only to suggest that the *moriscos* were like a "sierpe en el seno" [serpent in the bosom—my translation] of the body politic (Larsen "Sierpe"). Ricote seems to recognize that not all of his people were guilty as charged. Indeed, some of the *moriscos* were "cristianos firmes y verdaderos, pero que eran tan pocos, que no se podían oponer a los que no lo eran." [solid and true Christians. But there were so few that they couldn't compare with those who weren't]. In this same vein, he avers that his wife and daughter, whom he's lost track of in his wanderings, are "católicas cristianas" [Catholic Christians]. He confesses that he's not been so faithful in the new faith, but still avers: "tengo más de cristiano que de moro, y ruego siempre a Dios me abra los ojos del entendimiento y me dé a conocer cómo le tengo de servir" [I'm more Christian than Moor, and I always pray to God to open the eyes of my understanding and make me see how I am to serve Him]. One could debate the tone and temper of such assertions, but it is undeniable how much Ricote misses his home. His poignant "Doquiera que estamos lloramos por España, que, en fin, nacimos en ella y es nuestra patria natural" (pt. 2, ch. 54; 1071-74) [Wherever we are, we weep for Spain, where we were born, after all, and it's our native country] (749-50). Ethnic differences may divide, though Sancho and Ricote, at considerable risk, find ways to cross the chasm created between them. Their mutual reality will obviate *Realpolitik.* Their neighborly history together, a function of their mutual humanity, rewrites the Inquisitorial natural history that finally is so unnatural.

In Hispanic studies the debate continues as to whether there is an identifiably "Jewish writing," complete with indelible lexical markers, just as does the essentialist position that a person always will be a Jew, albeit unconsciously or inadvertently, after accepting (or after his/her ancestors accepted) some other religious faith.[12] Such was the Inquisitorial stance, that there could be no escape from one's lineage. The question comes to mind, whether there can be genetic markers in art, as well as whether alleles can have an aesthetic, not to mention a religion, as my friend and colleague, Seth Ward, puts it. Such essentialism is nothing new: the notion of what constitutes a convert, of what is the nature of conversion, and to what degree assimilation—or as seen from another perspective, absolute apostasy—is possible, would preoccupy not only the Inquisitors, but stretches back into Biblical times, as well as forward into our own. From the exploits of the Hebrew matriarchs, Sarah, Rebecca, Leah, and Rachel, who were converts to their husbands' religious practice, though occasionally presenting vestiges of the old ways—as Rachel's theft of her father Laban's *teraphim* [household gods] (Genesis 31:30-36)—to Saul of Tarsus (soon to become St. Paul), kicking against the goad and consenting to the martyrdom of St. Steven (Acts 8-9), the Judeo-Christian Scripture is replete with difficult and desultory conversions.

Even after visionary manifestations, belief and behavior modification become a process of evolution, albeit teleological or intelligently-designed. Moreover, rabbis from the fifteenth century (and even before), down to our own, have recognized that not all *conversos* were *anusim [forced ones]*, but that some were *meshumadim* [*willing apostates]*. The controversy continues (Gitlitz, Berg *et al*). At this point of my argument, I'll mention that on occasion I attend and lecture at meetings of the Society for Crypto-Judaic Studies. A salient feature of that particular gathering since its inception involves "return narratives," in which people describe having (re)discovered their Jewish heritage, in some cases more than five hundred years after ancestors were converted to another faith. These speakers describe residual practices and dimly understood performances, handed down from generation to generation. In many instances, they decide that they are Jewish, concluding in favor of what Francisco Márquez Villanueva terms "el judaismo de sangre" [Judaism of blood] (*Letra viva* 151). With reference to the individual genome, it is not so much that ontogeny recapitulates phylogeny, but rather that phylogeny dictates ontogeny (Gould, *Ontogeny*).

Such excavation of ancestral roots, not to mention their transplantation into the present, which may invoke some sort of (re)conversion, must remain basic human rights. At any point in one's life, even after many generations of absence, a Prodigal should be able elect to come home, becoming a *ba'al teshuva*, a "returned" and fully observant Jew. Occasional rabbinic rulings to the contrary, the Right of Return, whether to the Faith of the Fathers (and Mothers) or even to Israel itself, remains theoretically in force for all Jews, providing they can demonstrate an appropriate matrilineal Judaic heritage (Copans). Notwithstanding any nuances of the politics of identity currently in vogue, one wonders whether Judaism is somehow genetic and whether it or any other religion can be inherited. Such notions of physicality were basic to the ongoing operation of the Inquisition and the *Estatutos de Limpieza de Sangre*, as well as myriad forms of anti-Semitism. Writers have gone so far as attempts to isolate a person's political persuasion as a genetic function. Nature and nurture are notoriously difficult to separate: what one learns at one's mother's knee or around the dinner table, as opposed to what one brings from the womb are, in a human character, probably mutually indivisible (O'Donnell 34-39). One's identity need not necessarily be fixed, and may well continue fluid, even in permanent or semi-permanent flux. The very archetype of this is the Manchegan *hidalgo* who, rejecting genealogical constraints, refuses to be bound by conventions of whatever stripe. Such liberty to decide one's own destiny, despite an authoritarian atmosphere or antecedents, is hypothetically the bedrock of our American system. It almost goes without saying how radical such thinking was in Early Modern Spain.

Piqued and perplexed by what he perceived to be a pattern of persecution and paranoia, Américo Castro set off a firestorm, asserting that some Inquisitorial methods and modalities, particularly the preoccupation with lineage and whether *conversos* could ever become Catholics, might have their origin in the Synagogue (*Realidad* 31-32, 54 *et passim*). To my mind, Castro's reasoning goes beyond merely blaming the victims. I also recall a conversation with Thomas F. Glick at the very first meeting of the Society for Literature and Science in Worcester, MA, in the fall of 1987. He posited Castro's "Neo Lamarckism," whereby that venerable scholar understood that for the Inquisition, as well as the Synagogue, "acquired characteristics," including ethnicity, could be inherited from ancestors by succeeding generations. In turn, Yosef Hayim Yerushalmi writes concerning Freud's "psycho-Lamarckism," that is, his "belief" that cultural characteristics could be "inscribed" into a people's psyche, even into their flesh. A tradition is "encoded within the genetic legacy of a particular group," to be expressed "somatically" from generation to generation. Thus, "[t]he character traits embedded in the Jewish psyche . . . no longer require religion in order to be sustained," so "Jewishness is both inherited and indelible" (29-34, 52, 87-90; see also: Said 30-31, 46-48).

With this same Lamarckian bent, Yerushalmi quotes a Sephardic predecessor of Cervantes and his contemporaries, Solomon Ibn Verga, who in 1497 penned *Shevet Yehudah [Scepter of Judah]*, in which he argues against the forced conversion of his people, given that they are Jewish by nature:

> in the innermost chambers of their imagination they are Jews in every sense . . . My reason is as follows: The Jews who, when their religion was given to them, witnessed those awesome and holy assemblies and those marvels and fires from heaven, had the image of all that so strongly impressed upon them that it remains with them naturally. (qtd. in Yerushalmi 32-33).

Furthermore, Yerushalmi mentions a rabbinic legend (one that both Ibn Verga and Freud may unconsciously reformulate) that all Jews, those in the flesh at the time and those yet unborn, were "present at Sinai" to take on the "yoke of the Law" (34). One can debate the biological bases, not to mention the psychic parameters, for this sort of a priori implantation of racial or ethnic memories, but in certain instances, the natural history of human nature is not written without them. In a sense, this imagery continues the age-old psychosomatic debates as to how the spirit (nowadays, the mind) and the flesh might interact, influencing each other. One way or the other, the natural history of the spirit (or soul) has not taken place in a vacuum, nor has the natural history of Spain and Spaniards, all of them in the many-splendored varieties. Just in passing, this situation recalls to my mind aspects of the "naturalismo espiritual," as elaborated by Benito Pérez Galdós in another novel of a profoundly Darwinian, not to mention Cervantine, bent, his *magnum opus*, *Fortunata y Jacinta* (1886-87).[13]

Along much the same trajectory in the history of the United States of America, we encounter constructs parallel to those in Spain and Spanish domains. The forced expulsions of native sons and daughters, the Sephardim and the *moriscos*, recall our own sad story of dealings with Cherokee and Nisei, with Mormons and Muslims. As to racial policies, whether in the Southland or farther North, during and even long after Reconstruction and the heyday of Jim Crow, the One Drop Rule held sway (cf. Martínez). In other words, if a person had the minutest portion, no more than "one drop" of African (Negroid) blood, he or she was deemed African. The minutest amount of "black blood," the remotest "colored" ancestry or extraction, made one "Black." Now, here in America, the land of opportunity for all where the streets are proverbially paved with gold, just as in Golden Age Spain, many of "impure" blood were able to "pass," living life with a secret identity (Fuchs). Take for example the "hidden life" of the renowned literary critic Anatole Broyard (who had some African ancestry), as narrated by his daughter, Bliss Broyard in *One Drop* (Boston: Back Bay Books, 2008). In Early Modern Spain, as in twentieth-century America, such subterfuge would make for an existence of fear and trembling. In Iberia, practically everyone lived in mortal dread that their neighbors might discover the poisonous toad at the bottom of their well, or more in keeping with the imagery at hand, the key on the leather strap in their otherwise "pure" vintage. Themes such as *saber vidas ajenas* (*knowing other people's affairs*) became stock-in-trade of Early Modern discourse in Iberian and Ibero-American cultures (Silverman 197-212). Tomes such as the *Libro verde de Aragón*, a genealogical record compiled by the Inquisition beginning in 1507, enumerated the impurities of the Aragonese nobility and commoners alike, to the undoing of careers, even of lives. Curiously, Fernando II of Aragón, the so-called "Catholic Monarch," had a Jewish grandmother. For that matter, there's good evidence that his regal spouse, Isabel la Católica, Queen of Castile, also had Jewish ancestry. Not to mention Tomé de Torquemada, the first Grand Inquisitor, among other Spaniards great and small.

The notion of the melting pot which we hold so dear, here in America, Anglo-Saxon, as well as Latin, as in Golden Age Spain, often flares into an *auto-de-fe*. As a very current example, let us take former Senator, and now President, Obama's rise to prominence in a time when the politics of identity have only changed their dimensions, though hardly their significance. In turn, Jews have been considered, even in relatively recent American—Latin, as in Anglo-Saxon—society to be "black," though with appropriate ssimilation and accommodation, they can become "white folks" (Brodkin; Goldstein; cf. Martínez). As a young man, half African, half Caucasian, Barry Obama chose to be African-American, that is, Black (*Dreams from My Father*). Does his success with White voters, as well as with Black ones, give cause to feel more sanguine concerning the future? Or perhaps less so, given all the attention still paid to issues of blood? Time will tell. It seems that for many who went to the polls, economic issues trumped racial ones. But even in the Land of the Free, ethnicity remains a primary concern. In a very real sense, the President is still "passing," as no one is "pure" as to his/her racial and ethic make-up. In fact, in Obama's first post-election press conference, describing the purebred, hypoallergenic parameters for selection of a family dog, he referred to himself a "mutt." I also recall vividly that early on in his campaign there were questions as to whether he was "Black" enough.

Whatever *purity* means seems to vary with the times and the seasons, but in principle such concerns remain a constant. So it is—or has been—in Iberia, whether with Old Christians, with New Christians, and for that matter, with Jews. But it also seems that what so many of us construe as the American Dream, that no matter what the circumstances of one's birth, a person can legitimately aspire to, and, with the right combination of luck and effort, achieve any pinnacle. Moreover, Barack Hussein Obama is more or less a convert to Christianity, baptized into, as we have learned, a relatively militant Protestant Church on Chicago's South Side. His father's father had been a convert to Islam; his father was an adherent to that faith when he married the future Barack's Christian mother. Also a Muslim was his Indonesian stepfather, his mother's second spouse after her first union ended in divorce. As I wrote the first drafts of this essay, various Christian and Muslim authorities debated the veracity of his (and his father's and grandfather's) conversion/apostasy, arguing whether or not such life-changing transformations are possible, in his case specifically, as in more general terms (Hoyt). Obama's apparent calm, even serenity, is astounding, as a veritable maelstrom has swirled around him, just as it did around the *judeoconversos* and the *moriscos* in Early Modern Iberia, as their individual and collective nature continued to be debated. The debate concerning identity continues for all of us, as human life, individual and collective, continues to evolve. In short, the branch of natural history from which human life and lives continue to grow may best be construed as an evolutionary continuum.

Notes

1 A much-abbreviated version of this paper was presented at the Tenth Working Session of the Ometeca Institute, held in June of 2008, on the St. Petersburg campus of the University of South Florida. I'm grateful to all the participants in those meetings for their commentary, as I am to all who've read or listened to it since.

2 Castro cites the *Indice expurgatorio* of Cardinal Zapata (Seville, 1632), that a passage from chapter 36 of the 1615 *Quijote* was to be eliminated (*Hacia Cervantes* 213-21).

3 With regard to Cervantes and the *conversos*, see, among others: Castro, *Casticismos* 73-80, 130-43 *et passim*: Márquez Villanueva, *Letra viva* 151-68; Lokos 115-33; Finello 24-46; Yovel 276-83. Concerning Don Quijote's (re)fashioning of himself, see: Castro, *Casticismos* 17-143; Avalle-Arce 213-43; Silverman 204-12.

4 Concerning the import of humoral medicine in *Don Quijote*: see Juan Huarte de San Juan's *Examen de ingenios* (1575). See also: Soufas 1-36 *et passim*; Mariscal 48-49 *et passim*. Hereafter, references to *Don Quijote* in the original Spanish will generally be noted parenthetically according to part and chapter, as well as page number(s) in the Rico edition for direct quotes. The English renderings, unless otherwise noted, come from the Lathrop translation and will also be noted parenthetically in the text of the essay.

5 Regarding such apparent social contradictions, see: *Realidad* 94, 407, 432, *Casticismos* 32-38; Silverman 205-6; Shepard 100; Salazar Rincón; Finello 30-32; Yovel 280.

6 Hume 235-48. See also: Sugg, Jr. 96-102; Cohen 270-89; Sverdlik 69-76; Marshall 323-43; Paulson 83-89, 214; Gigante 57-58.

7 *Historia verdadera* and similar expressions appear throughout the novel, beginning in chapter 1 of part 1, giving rise to much critical investigation as to how (or not!) a work of fiction might be a "true history." See Wardropper 80-94; Nelson.

8 Concerning the growth of biological science as a discipline, see Mayr. Gould's wide-ranging "reflections on natural history" (the subtitle of *Ever Since Darwin*) generally look "forward," that is, after Darwin, though without neglecting the centuries and salient figures preceding him. Other volumes in which time's arrow points forward, with the study of Darwin's influence on contemporaries and future figures, include: *Darwin en España* (Barcelona: Península, 1982), by Thomas F. Glick, and *The Reception of Darwinism in the Iberian World* (Dordrecht/Boston/London: Kluwer, 1999), edited by Thomas F. Glick, Miguel Angel Puig-Samper, and Rosaura Ruiz. Other orientations are possible, however. Two of the leading lights in what is called "Darwinian" or "evolutionary" literary criticism figure in this present volume: the scholarship of David P. Barash and Joseph Carroll focuses on works and authors from long before Darwin, not to mention his contemporaries and those who came thereafter. With regard to the importance of natural history for Cervantes, his contemporaries, and the Iberian milieu in general, see: Garrote Pérez; Fernández Pérez 133-61; Pardos 117-40; Reed 167-84; Mariscal 42, 53-61.

9 Relative to questions of blood and *honra*, see: Castro, *Edad* 47-187, *Pensamiento* 355-69; Finello 24-46; Salazar Rincón 228-87; Yovel 263 *et passim*. Concerning *El retablo*, *Los alcaldes*, and *limpieza de sangre*, see:

Márquez Villanueva, *Letra viva* 151-52, 330-31; Silverman 206-7; Finello 29-32.

10 See: Shepard 86-89; Fajardo 191-205; Johnson 235-42; Molho 57-79; Sander Gilman 168-93; Geller, "(G)nos(e)ology" 243-82 and "Glance"427-44; Schrank 18-42; Anderson Imbert 69-73.

11 See: Epstein; Márquez Villanueva, "Planteamiento" 7-25, "Literatura bufonesca" 497-528; *España* 174-201.

12 See: Castro, *Realidad* 411-15; Stephen Gilman; Kaplan; Gitlitz, *Secrecy*; Wyschogrod; Nepaulsingh; Surtz 547-56; Round 557-76; Jacobs; García Cárcel 597-615.

13 Among those discussing Lamarck's influence on Darwin and Freud, along side of the impression of Darwin and Darwinism on Freud, see: Sulloway; Ritvo; Edelheit 45-74. Relative to Cervantes, Darwin, and Galdós, see: Larsen, *Tales*.

Works Cited

Alcalá, Angel. "Tres cuestiones en busca de respuesta: invalidez del bautismo 'forzado', 'conversión' de judíos, trato 'cristiano' al converso," in *Judíos. Sefarditas. Conversos.* Valladolid: Ámbito, 1995: 523-44.

Anderson Imbert, Enrique. "El tema de la nariz." *Boletín de la Academía Argentina de Letras* 62 (1997): 69-73.

Anonymous. *Lazarillo de Tormes*. Francisco Rico, Ed. Madrid: Cátedra, 1998.

Avalle-Arce, Juan Bautista. *Nuevos deslindes cervantinos*. Barcelona: Ariel, 1975.

Barash, David P. and Nanelle R. Barash. *Madame Bovary's Ovaries*. New York: Delacorte, 2005.

Brodkin, Karen. *How Jews Became White Folks and What That Says About Race in America*. New Brunswick, NJ: Rutgers UP, 1999.

Castro, Américo. *Cervantes y los casticismos españoles*. Madrid: Alianza/ Alfaguara, 1974.

---. *De la edad conflictiva*. 4th Ed. Madrid: Taurus, 1976.

---. *Hacia Cervantes*. Madrid: Taurus, 1967.

---. *El pensamiento de Cervantes*. Julio Rodríguez-Puértolas, Ed. Barcelona/ Madrid: Noguer, 1973.

---. *La realidad historica de España*. 6th Ed. Mexico City: Porrúa, 1975.

Carroll, Joseph. *Literary Darwinism*. New York: Routledge, 2004.

de Cervantes Saavedra, Miguel. *Don Quijote de la Mancha*. Francisco Rico, Ed. Barcelona: Instituto Cervantes/Crítica, 1998.

---. *Don Quixote*. Tom Lathrop, Trans. Newark, DE: European Masterpieces, 2005, 2007.

---. *Eight Interludes*. Dawn L. Smith, Trans./Ed. London/Rutland, VT: J.M.Dent/Charles E. Tuttle, 1996.

---. *Entremeses*. Miguel Herrero García, Ed. Madrid: Espasa-Calpe, 1966.

---. *Novelas ejemplares*. Francisco Rodríguez Marín, Ed. 2 Vols. Madrid: Espasa Calpe, 1962.

Close, Anthony. *The Romantic Approach to* Don Quixote. Cambridge/New York: Cambridge UP, 1977.

Cohen, Ralph. "David Hume's Experimental Method and the Theory of Taste." *ELH* 25 (1958): 270-89.

Copans, Laurie. "Rabbis' ruling puts thousands of converts in limbo." http://news.yahoo.com/s/ap/20081101/ap_on_re_mi_ea/ml_israel_conversion_crisis/print (accessed 11/5/08).

Edelheit, Henry. "On the Biology of Language: Darwinian/Lamarckian Homology in Human Inheritance," in *Psychiatry and the Humanities*. New Haven/London: Yale UP, 1978: 3: 45-74.

El Saffar, Ruth. *Beyond Fiction*. Berkeley/Los Angeles: U California P, 1984.

Epstein, Lawrence J. *The Haunted Smile*. New York: PublicAffairs, 2001.

Fajardo, Salvador J. "Don Quijote Wins by a Nose." *Hispanic Review* 70 (2002): 191-205.

Fernández Pérez, Joaquín. "La historia natural y la vida cotidiana," in *Felipe II, la ciencia y la técnica*. Enrique Martínez Ruiz, Ed. Madrid: ACTAS, 1999: 133-61.

Finello, Dominick. *Cervantes: Essays on Social and Literary Polemics*. London: Tamesis, 1998.

Fuchs, Barbara. *Passing for Spain*. Urbana/Chicago: U Illinois P, 2003.

García Cárcel, Ricardo. "Las mujeres conversas en el siglo XVI," in *Historia de las mujeres en Occidente*. Madrid: Taurus, 1992: 3: 597-615.

Garrote Pérez, Francisco. *La naturaleza en el pensamiento de Cervantes*. Salamanca: U Salamanca, 1979.

Geller, Jay. "'A Glance at the Nose': Freud's Inscription of Jewish Difference." *American Imago* 49 (1992): 427-44.

---. "(G)nos(e)ology: The Cultural Construction of the Other," in *People of the Body*. Albany: State U New York P, 1992: 243-82.

Gigante, Denise. *Taste: A Literary History*. New Haven: Yale UP, 2005.

Gilman, Sander. *The Jew's Body*. New York/London: Routledge, 1991.

Gilman, Stephen. "A Generation of *Conversos*." *Romance Philology* 33 (1979): 87-101.

Gitlitz, David M. "Hybrid Conversos in the 'Libro llamado el Alboraique'." *HispanicReview* 60 (1992): 1-17.

---. *Secrecy and Deceit*. Philadelphia/Jerusalem: Jewish Publication Society, 1996.

Gitlitz, David M. and Linda Kay Davidson. *A Drizzle of Honey*. New York: St. Martin's Press, 1999.

---, Irwin Berg *et al.* "Meshumadim: a discussion."http://www.kulanu.org/old/meshumadim.html (accessed 10/31/2007).

Goldstein, Eric L. *The Price of Whiteness*. Princeton: Princeton UP, 2007.

Gould, Stephen Jay. *Ever Since Darwin*. New York/London: W.W. Norton, 1977.

---. *Ontogeny and Phylogeny*. Cambridge/London: Harvard UP, 1977.

Hoyt, Clark. "Entitled to Their Opinions, Yes. But Their Facts?" "The Public Editor." *The New York Times* (1 June 2008). nytimes.com (accessed 20 Jan. 2009).

Huet, Marie-Hélène. *Monstrous Imagination*. Cambridge: Harvard UP, 1993.

Hume, David. Essay XXIII "Of the Standard of Taste," in *Essays Moral, Political, and Literary*. Indianapolis: Liberty*Classics*, 1985.

Hyman, Stanley Edgar. *The Tangled Bank: Darwin, Marx, Frazer and Freud as Imaginative Writers*. New York: Atheneum, 1962.

Jacobs, Janet Liebman. *Hidden Heritage*. Berkeley/Los Angeles: U California P, 2002.

Janson, H.W. *Apes and Ape Lore in the Middle Ages and in the Renaissance*. London: Warburg Institute, 1952.

Johnson, Carroll B. "Sobre la semántica de la nasalidad: El poema gongorino 'Contra una roma'," in *Jewish Culture and the Hispanic World. Essays in Memory of Joseph H. Silverman*. Newark, DE: Juan de la Cuesta, 2001: 235-42.

Kaplan, Gregory B. *The Evolution of Converso Literature*. Gainesville/ Tallahassee: UP Florida, 2002.

Larsen, Kevin S. *Cervantes and Galdós in* Fortunata y Jacinta—*Tales of Impertinent Curiosity*. Lewiston/Queenston/Lampeter: The Edwin Mellen Press, 1999.

--- "'Criar la sierpe en el seno': Sobre una figura que se enrosca y se desenrosca en las obras de Cervantes." Forthcoming in *Don Quixote: The First 400 Years* (Editorial San Marcos), Zenia Sacks DaSilva, Ed.

Lokos, Ellen."The Poetics of Identity and the Enigma of Cervantes's Genealogy." *Cervantes and His Postmodern Constituencies*. New York/London: Garland, 1999: 116-33.

Mariscal, George. *Contradictory Subjects*. Ithaca: Cornell UP, 1991.

Márquez Villanueva, Francisco. *Cervantes en letra viva*. Madrid: Reverso, 2005.

---. *De la España judeoconversa*. Barcelona: Bellaterra, 2006.

---. "Literatura bufonesca o del 'loco'." *NRFH* 34 (1985-86): 497-528

---. "Planteamiento de la literatura del 'loco' en España." *Sin Nombre* 10 (1980): 7-25.

Marasso, Arturo. *Cervantes. La invención del* Quijote. Buenos Aires: Hachette, 1954.

Marshall, David. "Arguing by Analogy: Hume's Standard of Taste." *Eighteenth-Century Studies* 28 (1995): 323-43.

Martínez María Elena. "The Black Blood of New Spain: *Limpieza de Sangre*, Racial Violence, and Gendered Power in Early Colonial Mexico." http://www.historycooperative.org/journals/wm/61.3/martinez.html (accessed 5 Feb. 2008). 38 pp.

Mayr, Ernst. *The Growth of Biological Thought*. Cambridge/London: Harvard UP, 1982.

Molho, Maurice. "Una cosmogonía antisemita: 'Érase un hombre a una nariz pegado'," in *Quevedo in Perspective*. Newark, DE: Juan de la Cuesta, 1982: 57-79.

Nelson, William. *Fact or Fiction: The Dilemma of the Renaissance Storyteller*. Cambridge: Harvard UP, 1973.

Nepaulsingh, Colbert I. *Apples of Gold in Filigrees of Silver*. New York/London: Holmes and Meier, 1995.

Obama, Barack. *Dreams from My Father*. New York: Crown Publishing, 1995, 2004.

O'Donnell, Erin. "Twigs Bent Left or Right." *Harvard Magazine* 108:3 (Jan.-Feb., 2006): 34-39.

Ortega y Gasset, Julio. *Meditaciones del Quijote*, in *Obras completas*. 3rd Ed. Madrid: *Revista de Occidente*, 1953: 1: 309-400.

---. *Meditations on Quixote*. Evelyn Rugg and Diego Marín, Trans. Urbana/ Chicago: U Illinois P, 2000.

Pardos, Fernando. "De Rocinante al rinoceronte: la historia natural y *El Quijote*," in *La ciencia y* El Quijote. Barcelona: Crítica, 2005: 117-40.

Paulson, Ronald. *Don Quixote in England*. Baltimore: Johns Hopkins UP, 1998.

Redondo, Augustín. *Otra manera de leer el* Quijote. Madrid: Castalia, 1997.

Reed, Cory. "Scientific and Technological Imagery in *Don Quijote*," in Don Quijote *Across Four Centuries*. Newark, DE: Juan de la Cuesta, 2005: 167-84.

Reston, Jr., James. *Dogs of God*. New York: Doubleday, 2005.

del Río Parra. Elena. *Una era de monstruos*. Madrid/Frankfurt am Main: Iberoamericana/Vervuert, 2003.

Ritvo, Lucille B. *Darwin's Influence on Freud*. New Haven/London: Yale UP, 1990.

Round, Nicholas Grenvile. "La 'peculiaridad literaria' de los conversos: ¿unicornio o 'snark'?" *Judíos. Sefarditas. Conversos. Op. cit.*: 557-76.

Salazar Rincón, Javier. *El mundo social del* Quijote. Madrid: Gredos, 1986.

Schrank, Bernice. "'Cutting Off your Nose to Spite Your Race": Jewish Stereotypes, Media Images, Cultural Hybridity." *Shofar* 25 (2007): 18-42.

Shepard, Sanford. *Lost Lexicon*. Miami: Ediciones Universal, 1982.

Sicroff, Albert A. *Les Controverses des statuts de "pureté de sang" en Espagne du XVe au XVIIe siècle*. Paris: Didier, 1960.

Silverman, Joseph H. "*Saber vidas ajenas*: un tema de vida y literatura y sus variantes cervantinas." *Papeles de Son Amadans* 89:267 (1978): 197-212.

Sugg, Jr., Redding S. "Hume's Search for the Key with the Leathern Thong." *TheJournal of Aesthetics and Art Criticism* 16 (1957): 96-102.

Sulloway, Frank. J. *Freud, Biologist of the Mind*. New York: Basic Books, 1979.

Soufas, Teresa Scott. *Melancholy and the Secular Mind in Spanish Golden Age Literature*. Columbia: U Missouri P, 1990.

Surtz, Ronald E. "Características principales de la literatura escrita por conversos:algunos problemas de definición," in *Judíos. Sefarditas. Conversos. Op. cit.*: 547-56.

Sverdlik, Steven. "Hume's Key and Aesthetic Rationality." *The Journal of Aesthetics and Art Criticism* 45 (1986): 69-76.

de Unamuno, Miguel. *Vida de Don Quijote y Sancho*. 4th Ed. Buenos Aires: Espasa-Calpe, 1943.

Varela-Portas de Orduña, Juan. "De la sangre vuelta vino (Notas sobre la ideología de la identidad en la primera parte de *Don Quijote*)." *Cervantes* 20 (2000): 79-100.

Wardropper, Bruce W. "*Don Quijote*: Story or History?" in *Critical Essays on Cervantes*. Boston: G.K. Hall, 1986: 80-94.

Wyschogrod, Michael. *The Body of Faith*. New York: Seabury Press, 1983.

Yerushalmi, Josef Hayim. *Freud's Moses: Judaism Terminable and Interminable*. New Haven/London: Yale UP, 1991.

Yovel, Yirmiyahu. *The Other Within*. Princeton: Princeton UP, 2009.

Yuval, Israel Jacob. *Two Nations in Your Womb*. Barbara Harshav and Jonathan Chipman, Trans. Berkeley/Los Angeles: U California P, 2006.

CHAPTER 8

The Selective Female: A Darwinian Interpretation of Aluísio Azevedo's *O cortiço*

Aluísio Azevedo's *O cortiço* (1890) is considered the Brazilian naturalist novel *par excellance* (Bosi; Candido; Pereira). The naturalist period, it should be noted, begins in Brazil in 1881with the publication of *O mulato*, also written by Azevedo. In naturalism, as is well known, characters and plots are guided by the blind destiny of the "natural laws" that the science of the time believed it had deciphered. Among these natural laws were those coming from Darwinism and other types of evolutionism.

The objective of this chapter is to analyze the relationships between gender and race in some of the characters in *O cortiço* from the point of view of parental investment. We do not believe, of course, that Azevedo had the concept of parental investment specifically in mind. However, the way the naturalist sensibility treats the human relationships in the novel demonstrates that it is not necessary to be conscious of the underlining structures for the writer to avail himself of them. The author of *O cortiço* does not need to be a psychologist to weave a plot in which men and women, both black and white, become involved in psychologically tense situations. And one does not need to be an evolutionary biologist to verify that there arise some very clear conflicts of interest in the novel that can be best understood as part of human nature. Obviously, as Gilberto Freyre says (607), Azevedo has a sociologist's intuition that makes him capable of detecting and describing the tense social relations that in Brazil both anticipated and followed the abolition of slavery and the proclamation of the republic. We believe, however, that Azevedo's Darwinism complements a sociological reading with important psychological and biological dimensions that enrich the reading, by pointing out conflicts that are not socially constructed, but rather rooted in our animal condition. Our animal nature does not become weaker with the advent of culture, but rather it modifies itself and takes on the forms that are studied by the sociologist and the anthropologist.

Just like every heir of the modernist generation of 1870, Aluísio Azevedo participated in the boundless belief in science and progress, in the desire to challenge and to topple archaic ideals based on religious dogma, and in the necessity of renewal of the national institutions following the European model (Sevcenko; Bosi). In his novels, themes such as social organization, differences between the genders, slavery, miscegenation, and prostitution are presented in a scientific form which is characteristic of the naturalist movement. In this form, one can see also the strongly conservative social criticism so easily observable in the elite of the eighteen hundreds. The struggle for survival, for instance, in a novel like *O cortiço* is clear and the message leaves no doubt: in order to win it is necessary to forget the morality imposed by decadent institutions that praise romantic love and heroic and selfless attitudes towards life and death.

One can observe that it was through the literature of the time, both novels and poetry, that the Brazilian elite tried to cultivate its own "instinct of nationality" (Candido; Schwarz; Skidmore), an expression made famous in the title of an essay by perhaps our most famous man of letters, Machado de Assis. In that essay, Machado criticizes those who think that to create a national literature, and with it the identity of a nation, it is necessary to present local color. But, in contrast to Shakespeare, who seeks many of his themes outside England, Machado fills his universal plots with Brazilian characters and background, which are actually very much based on the Fluminense (the region around Rio de Janeiro) culture, above all on its conflicts with the metropolitan ideals of the Portuguese Empire. In the same way, in *O cortiço* the relationship of the Portuguese with the Brazilians, of whites with the descendants of Africans, and of the foreigner with the native all have an important place in foregrounding the relations between miscegenation, degeneration, and social disintegration, all of which lead to the argument for the whitening of the Brazilian race as a necessary condition for the country to enter modernity. Written in a moment of great development in the biomedical sciences, and the simultaneous degradation of the health conditions of the country, Azevedo's book is part of a complex scientific and cultural scene where the mixture of widespread yellow fever, chicken pox, tuberculosis, cholera, and typhoid fever, coexist with the development of fields of biological investigation such as microbiology, immunology, parasitology, bacteriology, embryology, evolutionism and the nascent science of eugenics. Often, these different kinds of knowledge have a very special appeal in Aluísio Azevedo's work (Bosi; Sodré; Candido).

Two works published in Brazil in 1875—one scientific and one literary—make specific references to Darwinism: the Ph. D. dissertation by Augusto César de Miranda Azevedo, *Beribéri*, and the serialized novel by Emilio Zaluar, *Dr. Benignus*. These two texts can be interpreted as the embryo of a theoretical position that would be developed more and more in the following years: the mixture between Darwinism and systems of racial classification. Miranda Azevedo's work, *Beribéri*, is a Ph. D. dissertation, and Darwinism only appears as a question that one of the members of the dissertation committee proposed for the candidate to answer in writing, as was customary at that time. In response, Miranda Azevedo gave seven lectures about Darwinism in the "Public Lectures of Glória." Only one of these seven texts survived. In it, Miranda Azevedo defends Darwinism with almost the same words as Ernst Haeckel. As for Zaluar's literary work, it is clearly lacking in aesthetic value. Built within a schematic and naïve naturalism that borders upon mediocrity, the characters are absolutely flat: the good scientist who is wise and, naturally, white; the white colonist who is stupid and incorrigible; the savage native who could be rehabilitated by science and civilization. The reference to Darwinism is limited to the idea that the science of biology is necessary and sufficient for man to fulfill his reign on Earth. The idea of man as the apex of creation is manifest and there is nothing that one can detect about the intellectual revolutions that Darwin's ideas inspired in the human mind. From these humble beginnings, from 1881 on, during the whole "Naturalist episode" (Sodré), writers such as Aluísio Azevedo, Inglês de Souza, Adolfo Caminha, and Júlio Ribeiro, among others, focused on concepts such as the "struggle for survival," "survival of the fittest," or "the strongest," "adaptation," "moral evolution" and "progress."

Born in São Luis of Maranhão, Aluísio Azevedo, upon his brother Artur's invitation, moved to Rio de Janeiro when he was 19 years old. At the time, Artur was already a known cartoonist for newspapers. Azevedo returned to Maranhão in 1878, after his father's death, and started working in the press and investing in a writer's career. It was in his second book, *O mulato*, in which Aluísio Azevedo started following the lines of Naturalism. This novel ended up making him many enemies in São Luís, because it pictures the prejudices of the local society of the time. Upon his return to Rio de Janeiro, Azevedo starts trying to live from his writing, and thus becomes perhaps the first professional writer in Brazil. This career lasted 10 years. In 1891, he obtained a position in the government of the State, and drifted away from literature, to abandon it completely when he started a diplomatic career. Aluísio Azevedo actually sold the copyrights for all his literary work to H. Garnier, and finally died in Buenos Aires in 1913, when he was only 55 years old.

In his first novel, *O mulato* (1881), we see the beginnings of his evolutionist bent, as he struggles to escape the prevailing realism and romanticism of his day. In the novel the character Raimundo's father has died and he distances himself from his mother, a former slave, by traveling to Europe to study. He returns from Europe with a degree and goes to live with his uncle and tutor, Manuel Pescada, and ends up becoming interested in his daughter, Ana Rosa. She is also interested in him, but there are obstacles to the romance due to the restrictions imposed by the family because of the young man's black mother.

One of those most interested in preventing their union is the Canon Diogo. This man is the same who killed Raimundo's father when he discovered him having sex with his white wife, Quiteria. Raimundo, crushed after he is told he cannot have Ana Rosa because he is the son of a black woman, decides to go to Rio de Janeiro. Ana Rosa asks him to stay. They end up making love and Ana becomes pregnant.

The two intend to elope, but their plans are frustrated by Diogo and Dias, one of Manuel's clerks who wanted to marry Ana Rosa. When Raimundo returns to his house, he is shot dead by Dias. Ana Rosa, seeing him dead, has a miscarriage. Years later, however, she appears in an official reception, married to Dias, and worried about the three children they left sleeping at home.

In this first effort, it is possible to identify several realist elements in the plot, such as the social criticism, the anti-clericalism, the struggle against racial prejudice, the sexual aspect (present in the carnal nature of Ana Rosa's love for the mulatto), and the triumph of evil. There are, however, some characteristics of Romanticism as well, such as the idealization of the hero, the cliché plot of the love that is not fulfilled due to traditions and prejudices, and the extravagantly complicated nature of the story.

In *O cortiço (The Tenement House*, 1890), the plot is not organized around one hero, but rather around a whole society that is formed around the tenement house of the title. In this novel there are several very clear characteristics of Realism-Naturalism: determinism, with men shown as products of the environment; social criticism, with the exploitation of man by man; and zoomorphism, or the reduction of man to the animal level, acting according to instincts, seen in several figures of speech and in the super-valorization of sex.

In the novel, João Romão is an ambitious Portuguese who manages to open a small commercial establishment. He moves in with an escaped slave, Bertoleza, who is his neighbor, and with her money he increases his property. He then starts renting out little houses. The business is successful, and the cubicles multiply, thus becoming a tenement house. An enormous variety of human types start living in the place, working in the quarry started by João Romão, and buying groceries in his story. The only one who is upset with the growth of the tenement is another Portuguese man, Miranda, who belongs to a higher social class.

Among the types that live in the cortiço are Machona (Big Dyke), a loud-mouthed washerwoman, "whose children did not look like one another;" Alexandre, a pedantic mulatto; Pombinha (Little Dove), a fragile young woman who loses herself due to the influences of bad company; Rita Baiana, a beautiful mulatta who, in the beginning of the story, lives with Firmo, a rowdy mulatto; Jeronimo and his wife, and many others.

Another cortiço begins to take shape on the same street, and receives the nickname "Cabeça-de-Gato" (Cat's Head). The residents of Cabeça-de-Gato, in their turn, nickname the residents of the tenement house "Carapicus" (a local Brazilian name for a type of fish). There is rivalry among the inhabitants of the two cortiços, which is aggravated by the dispute between Jeronimo and Firmo for Rita Baiana. Firmo then moves to Cabeça-de-Gato. The conflict only ends after a mysterious fire destroys part of João Romão's cortiço.

The Portuguese, however, rebuilds the tenement house and makes it even more prosperous. His dream then becomes to marry a girl from a good family, his neighbor Miranda's daughter Zulmira, in order to increase his status. After the two Portuguese man come to an agreement on this marriage, João Romão decides to get rid of his old companion Bertoleza, in order to open the way to his wedding.

Because the objective of this chapter is to interpret *O cortiço* in evolutionary, Darwinian terms, we shall take as our points of departure the four concepts that Joseph Carroll presents in the beginning of *Evolution and literary theory*:

1) The relationship between the individual (or a species) and the environment in which it lives, considering that

the environment is constituted by many individuals of the same and of different species. Among the most important interactions to be studied are gender relationships. It is not necessary to emphasize how much such relationships are important from sociological, anthropological and ethnographic points of view;

2) The presence (in the characters, in the narrator, in the author, in the readers) of innate psychological structures, as well as perceptual, rational and effective structures that evolved through an adaptive process of natural selection. Such structures regulate, but do not blindly determine, the mental life (of the characters, of the narrator, of the author, of the readers). That is, we will try to find, among the behaviors of the characters, those that represent a psychological structure which is neither conscious nor social;
3) The interaction, in the human motives, among the near (proximate) or immediate causes, regulated by the principles of inclusive aptitude, and the distant (ultimate) causes. In other words, the explanations that we can give for any behavior or human action, in terms of functional or proximate, immediate causes must always be complemented by explanations of how, in historical (evolutionary) terms, such behaviors were developed. The differentiation between the immediate and the historical causes allows us to see that human behavior, just like our conflicts of interest, are regulated, although not blindly determined, by structures that are beyond any rational will.
4) The phenomenon of representation (falsely refuted by many postmodernists, poststructuralists, and above all by relativists), including the literary representation, recognized as a form of "cognitive map." That is, we will look at Aluísio Azevedo's work as an extension of the adaptive orientation of these human beings in an environment that is first of all spatial and physical, but also social, cultural and political. The Rio de Janeiro of Azevedo's novel paints an environment that, being historical in the sense that much has changed since then, is also biological in the sense that it represents the behavior of living biological beings, men and women, slaves and landowners, blacks and whites, who exhibit clear conflicts of interest, which are not

> merely class conflicts, or gender and race conflicts, but rather biological and genetic conflicts, although many of these are merged with the prejudices of the time.

In a more specific way, we will be using two categories that come from the works of biologist Robert Trivers: 1) parental investment; 2) conflict of biological interests. Trivers define parental investment as "...the investment of one of the parents in an individual child in order to increase this child's chances to survive [...] at the cost of investing in another child" (67). The difference between feminine and masculine behaviors, in political societies culturally stratified as ours, is one of the areas in which one can see the affirmation of the social, non-biological character of human behavior. In psychological terms, these behaviors are socially constructed. However, human reproduction is a biological phenomenon with a psychological impact that can hardly be overrated. Human reproduction depends on a long gestation period, *which occurs inside the woman's body*, and gives the man a morally scandalous freedom to decide whether to abandon the woman and the child. Hence, the parental investment of the female is disproportionally larger. This might perhaps be in some way incorporated in the law that usually gives the mother the guardianship of the child. Whether she wants it or not, in order to have a child, the mother lives through radical changes in her body and in her mind. Physically, the initial investment of the male, the sperm, enables the male to have innumerous children at the same time. The woman, unless she is pregnant with twins, has to wait at least nine months to have more than one baby. The other side of this matter is that only the mother can be sure whose the child in her womb is. There is a proverb in African culture that says "Mommy's baby, Daddy's maybe" (Buss) that reflects this uncertainty.

Trivers's theory of the asymmetry, "female *versus* male," of parental investment is explicitly based on the theory of the gene as a fundamental level of natural selection, incorrectly but popularly known as the theory of the selfish gene. Animals are altruistic principally when they defend their genetic representatives. Mammalian mothers invest much time and resources in the genes of their offspring, from the moment these install themselves in their womb, whereas the fathers initially invest little time and a few relatively cheap spermatozoon. Because of this the females *tend* to be more selective in their choice of partners for copulation. And, although there naturally are conflicts of genetic interests beyond the relationship male/female (one can think here about the relationships among fathers, mothers, daughters and sons, between brothers and sisters, between cousins, etc; and there is also the relationship among non-related individuals), such a relationship is a constitutive focus of our personality and psyche, since we are born either male or female, and we are also socialized in the sense that, in the future, we will have a stable relationship with a person of the opposite sex, with the explicit intention to constitute a family. Let us now see how this plays out in *O cortiço.*

We will start with a couple that appears in the beginning of the novel: Estela and Miranda. From this first couple, we will analyze other pairs, triangles, and assorted geometric figures formed by the characters in the novel. At the same time, we will be particularly attentive to the behavior of one male character, Jeronimo. The behavior of these characters demonstrates the shared interests and conflicts of interest among them and their partners. Such conflicts contribute to structure the plot of *O cortiço*, and part of Aluísio Azevedo's art resides in representing at least one character who changes during these conflicts. Although it is justly classified as a naturalist novel whose characters are psychologically constituted as fixed types, from the theory of types (Auerbach), Jeronimo's degeneration is going to point precisely to the possibility of change, the same way that we can see the change in his partner, who claims the right to "better" her race by mixture with Portuguese blood.

O cortiço is structured in twenty-three chapters or episodes numbered with Roman numbers I to XXIII. This analysis will start with a reading of the first two episodes, more specifically the conflict between Estela and Miranda.

In the first lines of the text we read how the Portuguese shop owner João Romão manages to purchase the store. It also shows his relationship with the black slave Bertoleza. They form an ambiguous pair that can be interpreted as the exemplary nucleus both of the gender and race relations: "He (João Romão) proposed they lived together and she agreed with open arms, happy being once again involved with a Portuguese because, as every black woman, Bertoleza did not want to subject herself to black men and instinctively sought a man from a race superior to hers" (8).

Later João Romão will plot to marry another character, this time a white women. But already in the first episode the text informs us that Miranda, a rich Portuguese businessman, bought the two-story house that was precisely on the lot João Romão coveted for his tenement house. Let's pay attention to the human reason that leads Miranda to live in the two-story house in Botafogo beach. Miranda explains to his colleagues that he was moving to that place "Because his wife, Dona Estela, a pretentious woman with airs of nobility, could not stand living downtown, and also because his girl, Zulmirinha, was growing very pale and needed space to become stronger and grow up (12). However, this was the explanation he gave to other people, since:

> The real cause for the move was in the need, which he recognized as urgent, to distance Dona Estela from the reach of his clerks. Dona Estela was a rascal of a little woman: she had been married for thirteen years and during this time had given the husband all kinds of troubles. Even before the end of the second year of the marriage, Miranda caught her in a flagrant act of adultery; he got furious and his first impulse was to send her to the devil with her accomplice; but his commercial house was guaranteed with the dowry she had brought, about eighty *contos* in buildings and stocks of the public debt, which

> the wretched man used as much as the dowry system allowed. (12)

We see then that the sexual jealousy and the feeling of possession are mixed up with economic interests in Miranda's spirit. It seems that what is happening is a kind of game or struggle between selfish people. Miranda does not separate from Estela, but both hate each other, although further ahead the male's sexual thirst will transform the couple's relationship into a "mixture of sexual happiness and moral repugnance" (15). And, what is worse, the conception of a child further complicates the dilemma:

> The birth of Zulmira aggravated the situation even more; the poor child, instead of being a way to connect the two wretches, was more a new isolating factor established between them. Estela loved her less than the maternal instinct asked because she supposed she was the husband's child, and he detested her because he was sure he was not her father. (15)

The fact that Estela loved her daughter less "than her maternal instinct asked" could be seen as a weakening of the thesis that the feminine parental investment is prominent when compared with the male's investment and that, psychologically, the mother's investment establishes a profound connection between her and her offspring. Because, independently from whom the father is, Zulmira carries Estela's genes, and up to the point the text enables us to see, Estela does not have other children. However, we cannot forget that Estela probably did not choose Miranda on her own. She too does not recognize in him a desirable male, as we can conclude by her adulterous behavior. It is probable that if she believed that Zulmira was the daughter of one of her lovers, she might love her more. Ironically, Estela only *supposes* that Zulmira is Miranda's child. Naturally, a loyal wife (or, principally nowadays, a very careful wife) can be *sure* who the father is. On the other hand, the conviction, never tested by Miranda (there were no DNA tests at the time) that Zulmira was not his daughter, especially because of his wife's adulterous past, can be interpreted as a typical feeling of a living being who belongs to a species in which the female's ovulation is invisible to the male, and that he actually never knows for sure if the child in the partner's womb is really his. This type of asymmetry in the relationship between the woman and the man is expressed in a behavior with various literary manifestations already considered classic. Be it in *Othello* or in *Dom Casmurro,* the characters of the stories are involved in a true prisoner's dilemma, because they are not sure about what goes on or went on in another person's mind, respectively their Desdemona and Capitu.

Parallels of this kind (different behavior between female and male, between biological mother and father) are found in many mammal species, including, for example, seals, gorillas, and lions. Among the sea elephants of Patagonia *(mirounga leonina),* a male dominates a crowd of females. A male of this species can reach six tons, while the female rarely goes beyond the nine hundred kilos. The enormous sexual dimorphism seems to lead the males to fight for the females. To the winner, everything; to the losers what remains is to keep hanging around the harem trying to find a break in the attention of the alpha male (MacCan et all)

Some groups of birds, on the contrary, behave in relation to the offspring in a different way, because the egg develops externally to both of the parents, and it depends on the care that only the two parents can offer. Among fish, there are species in which the females eat their own offspring if there are not defended by the fathers, who (the fathers) make nests of water bubbles, expel the female from his territory and spend the day carrying their babies in the mouth and bringing them back to the nest. There is also the sea horse, whose male keeps the babies in a sac. In these species, it is the females who tend to be polygamous (Vincent and Woodroffe). In the midst of all this diversity of sexual systems we can find also some cases of species, including of mammals, which are, at least for some time, monogamous. This seems to be related, in some cases, to the production of a hormone called oxitocina (love's molecule), and also with cellular receptors specific for this hormone. Hormones and receptors are naturally codified in the genes and are created and maintained through natural selection (Ridley).

This means, in the human species, that a woman's betrayal can lead a man to spend his resources on children that are not his. On the contrary, a man's betrayal—no matter how chauvinistic this may seem—does not lead necessarily to a compromising of the betraying male in the sense that he might bring in his body the baby of another female. Nature made us this way. But biology is not destiny. If the males tend to be more aggressive with the females because of their genes, this does not mean that we should be more tolerant with them about this natural aggressivity. On the contrary, knowing that the males have this propensity to become aggressive, our laws must be designed so that they can punish any behavior leading to violence. Miranda does not actually physically attack Estela; his financial interest surpasses his sexual jealousy.

There are other sexual triangles in *O cortiço* in which violence is consummated. The most evident is the triangle among Rita Baiana, Firmo, and Jeronimo. Rita Baiana was an attractive mulatta who lived in the tenement house, but in the beginning of the story she had disappeared with Firmo, a vagabond capoeira player, who was the same color as she. In this pair there will be the intrusion of Jeronimo, a married Portuguese man, who has a little daughter, and whose family came to Brazil to try to obtain a better life. Jeronimo, his wife, and daughter come to live in the tenement house, and he goes to work in the quarry located in the back of Romão's land. But though João Romão is the typical Portuguese who comes to Brazil and continues being Portuguese, Jeronimo is different, belonging to those whose personality is engaged by the Brazilian types like Rita Baiana.

Due to the strength of his dedication to the work, Jeronimo became as good as the best workers in the quarry, and reached a salary as high as theirs:

"within two years, he became so distinguished among his colleagues, that the boss transformed him into a kind of foreman and increased his salary to seventy thousand réis" (56). Jeronimo also had "the strength of a bull, and that made him respected and feared by the whole group of workers, as did also, perhaps, the great seriousness of his character and the austere purity of his habits" (58). He was a man of unquestionable honesty and of a primitive simplicity in his way of life. "His little daughter was always clean and well fed... in spite of the difficulties against which both [Jeronimo and his wife] fought in their first times in Brazil" (56).

In the middle of the episode VII the reader can follow the preparations and the development of two great parties that happened in the tenement house, one organized by Rita Baiana, and the other by another neighbor (das Dores). Both parties enraged Miranda because of the music and the screams that came from the tenement house: "Jeronimo and his wife were invited to both tables, but they didn't accept the invitation of any of them, since they were more disposed to spend the afternoon at each other's side, very quietly as usual, eating peacefully their traditional food and drinking their quarter of green wine from the same cup" (75).

However, after the dinners were served and the noise level went down, from the door of his little house in the tenement house Jeronimo starts playing a Portuguese guitar:"At this moment a guitar started moaning at the door of number 35; it was Jeronimo. After the rowdy joy and good humor in which the whole republic of the tenement house had rejoiced, his guitar seemed even sadder and more reminiscent of his country than ever before" (78). But soon, the Portuguese song is interrupted by a Brazilian rhythm:

> But, suddenly, Porfiro's bandolin, followed by Firmo's guitar, irrupted with a vibrant Bahian choro. Nothing more than the first notes of that criollo music was needed to make the blood of all those people wake up, as if somebody whipped their bodies with wild nettle. And

> other notes, and others, followed, each more fiery and more delirious. They were no longer two instruments that sounded; rather, they were lubricious moanings and sighs loose in a torrent, slithering like snakes in a burning forest; there were convulsive ahs, cried in a frenzy of love; it was a music made of kisses and delicious sobs; tenderness of beasts, tenderness that hurt, making one explode in pleasure. (78)

We can understand that the music is a preface to the fight that will happen soon afterwards. Firmo's music is going to silence Jeronimo's. His song dispels sadness, and makes the body dance. From this moment, a revolution begins to happen in Jeronimo, canalized by Brazilian music and by the dancing curves of Rita Baiana's body. The duel to obtain the mulatta involves not only Firmo, but also even Jeronimo's wife, Piedade de Jesus, who was Portuguese. In the next episode, VII, when Jeronimo misses one afternoon at his job, the apparition of Rita to help the sick man affects Piedade in a very profound way: "The Portuguese woman didn't say anything; she smiled in a forced way while, deep inside, she resented that invasion from that stranger to take care of her man. It was not intelligence nor reason that pointed out to her the danger, rather, it was instinct, a subtle and mistrusting sense all females have against the others, when they feel their nest is exposed" (77). The competition is clear, and before that, still in the episode VII, the reader certainly sees reasons why the "instinct" of Jeronomo's wife is not wrong. Jeronimo, like other residents of the tenement house, had been touched by the fiery music that "...flew madly in the air like a warm aroma of Brazilian plants, around which feed sensuous flies and poisonous beetles, who gyrate frenetically, drunk from the delicious perfume that kills them with voluptuousness" (78):

> Jeronimo distanced himself from his guitar and kept quiet, his hands forgotten on the chords, all attentive to that strange music, which came inside him to continue a revolution started since the first time he felt on his face—

> like a slap in a challenge for a duel—the light of this proud and savage sun, and his ear heard the refrain of the first cicada, and the first fruit he tasted in these fiery lands spread its sweet and acidic taste in his mouth; and his soul felt the tenderness of the aroma of the first bogari, and his blood went crazy with the smell of the first woman, the first mestiza who shook her skirts and her hair near him. (79)

And, to his wife's unhappiness, Jeronimo:

> Saw Rita Baiana, who had gone inside to exchange the dress for a skirt, coming outside with naked shoulders and arms, ready to dance. The moon appeared in the sky at that moment, surrounding her with silver rays, and intensifying the mulatta's movements, which were full of an irresistible, simple, primitive grace made completely of sin, full of paradise, with much of a serpent and much of a woman. (79)

> She jumped in the middle of the circle, arms on her waist, moving her hips and circling her head, sometimes to the left, sometimes to the right, as if in a pain of carnal pleasure, in a luxuriant movement that took her breath away; sometimes she ran with her belly raised, then she stepped back with her arms extended, trembling, as if she was sinking in a pleasure thick as oil, a space where there is no end. Then, as if returning to life, she exhaled a long moaning sound, snapping her fingers in the air and bending her legs, lowering herself, standing up again, without ever stopping the movement of her hips, and then suddenly she tap-danced, small steps, frenetically, lifting and lowering her arms which she bent, one and the other, over the nape of her neck, while her flesh boiled fiber by fiber, in great tension. (80)

Firmo is between Jeronimo and Rita Baiana. The resolution to this conflict, the details of which I won't reveal here, revolves around the Portuguese's violent victory. Without knowing what had happened, Rita Baiana had already decided for Jeronimo. She too had her reasons:

> Since Jeronimo became interested in her, fascinating her with his tranquil seriousness of a good and strong animal, the blood of the mulatta demanded its rights to purify itself, and Rita preferred in the European the male of a superior race. The digger, on his turn, giving in to the instinctual impositions, became tired of his wife, who was from his own country, and wanted the mulatta, because the mulatta was pleasure, was voluptuousness, was the golden bitter fruit of these American land, where Jeronimo's soul learned the lust of the monkey, and where his body sweated the sensual smell of the goats. They made love brutally, and both knew it. This irrational and empirical love had become even more charged, for both of them, since the tragic incident of the fight, in which the Portuguese was the victim. In her eyes, Jeronimo acquired the aura of a martyr sacrificed to the loved woman; her love grew with that blade that attacked Jeronimo, it lighted up with the spilled blood, and, afterwards, her absence in the hospital completed the crystallization of his prestige, as if the digger had gone to his grave, dragging behind him the longing from all those who cried. (80)

In this passage the "instinctual impositions" evoke the environmental determinism of the inclement Brazilian climate which forces Jeronimo to change. But, alongside this one—and this is what most interests us—since it points to an ethnic assimilation seen favorably by the character herself: "Rita preferred in the European the male of superior race." In the very beginning of the novel we see that Bertoleza "did not want to subject herself to black men, and instinctively sought a man in a race superior to hers," and that is why she decided to live with Romão. But Romão, when he no longer needs her, will leave the black slave. Jeronimo does not do the same: he is another kind of male. He was taken over by the music, by the woman, by the primitive Brazilian sun. Such primitivism, such inferiority, would fit very well in some evolutionist schemes and, indeed, literary historiography points out these possible readings. However, it is a common error to think that Darwinism agrees with the dualism of "superiority *versus* inferiority." Additionally, the evolutionism of Aluísio Azevedo cannot be considered Darwinian. Perhaps it is Spencerian or Haeckelian, or a mixture of both. The point is that Rita Baiana chooses Jeronimo because he is strong, and his past dedication to the wife and child also mean that he is trustworthy. Jeronimo is enraptured by the Brazilian land, a force superior to his own, but he maintains his character as good husband and good father.

Rita Baiana's choice for Jeronimo synthesizes two ideas of sexual selection, both well known among today's evolutionists. Jeronimo is strong, therefore has good genes; but he is also a dedicated father, always willing to invest his resources in his daughter. Rita Baiana chooses Jeronimo because of is physical strength, his success as a provider, and also because of his past reputation.

At least since the publication of *Descent of man and sexual selection*, in 1871, sexual selection has become a process that could or should contribute to the explanation of some animal behaviors considered bizarre, such as the classic example of the peacock's tale. The perfecting of the male is meant to please the female. Wallace considered sexual selection a particular case of natural selection. Darwin saw, for instance, in the size and form of the horns of the Irish elk evidence against natural selection, because such a rack did not contribute, and could even obstruct, the animal's survival. There must be another process going on, namely the increased attention, on the part of the females, to the more showy antlers. An interpretation for this selective behavior by these females is that the antlers are an expression the male's general health. Animals suffering from worms have asymmetry and bad formations in their visible structures. The ones with bigger antlers would attract on average more females and leave more numerous offspring. Additionally, signs of the male's involvement, of parental investment, seem to provoke a special attraction over the females of other groups of animals.

As we saw, Jeronimo is strong and involved. But he is also white and male. The key structured by Aluísio Azevedo presents the white male as superior.

And Rita herself recognizes her right to social ascent. Does this mean that Aluísio Azevedo was/is a racist? Beyond the anachronism of the question, it is necessary to consider that all this naturalist rawness does not hide an ironic criticism of how society sees the differences between males and females, between blacks and whites. This idea of superiority, especially of the white individual over the darker one, of the Portuguese over the locals, is identified many times, mainly among the Brazilian scholars, as Darwinism.

Azevedo is a human being and represents his characters as general types (the Portuguese is rich, the businessman who becomes rich through usury, the black slave woman who is hardworking and honest, the capoeira dancer who is mulatto and a rascal) at a time in which the importation of literary, scientific, and philosophic ideas was the norm in Brazil. French Naturalism and Realism, as well as the same movements in Portugal, were, in stylistic terms, going to meet with Darwinism and other forms of evolutionism at the end of the Nineteenth century. Although in *O cortiço* neither Spencer's social Darwinism, nor Haeckel's Recapitualism appears explicitly, the concept of struggle and its related concepts appear many times in the text: the word "struggle" appears 14 times; "war" 12 times; "dispute" 5 times, and "jealousy" appears 9 times. It seems clear that what it at stake, since this is a naturalist work, is the presentation of an atmosphere in which the characters are represented in their most brutal aspect.

Jeronimo, however, is perhaps a symbol of transformation, although here the Portuguese man is assimilated by the local element, characterized as primitive. On the other hand, from the point of view of Rita Baiana herself, what happens is a real mixture that is going to improve the mulatta's offspring, thus whitening her. It is curious to observe that from Jeronimo's point of view his relationship with the mulatta results in a degeneration of the white race; however, from Rita Baiana's perspective, the mixture is an improvement, and that is what the Brazilian elite of the time praised

Works Cited

Auerbach, Erich. Mimesis: a representação da realidade na literatura ocidental. São Paulo: Perpectiva, 2004.

Azevedo, Aluísio. *O cortiço*. Rio de Janeiro: Record [1890], 1995.

Azevedo, Miranda A.C. "Darwinismo". *Conferências Populares da Freguesia da Glória*. Rio de Janeiro, Typographia e Impr. Const. de J. Villeneuve S. C., n° 1 a 10, 1876.

Bosi, Alfredo. *História concisa da literatura brasileira*. São Paulo: Cultrix, 1995.

Buss, David. *A paixão perigosa*. Rio de Janeiro: Objetiva, 2000

Cândido, Antonio. *Formação da literatura brasileira.* 7th ed. 2 vols. Belo Horizonte: Itatiaia, 1993.

Carroll, Joseph. *Evolution and Literary Theory*. Missouri: U of Missouri P, 1995.

Darwin, Charles. *Origem das espécies*. São Paulo: Edusp, 1985.

Freyre, Gilberto. *Sobrados e Mucambos.* 7th ed. Rio de Janeiro: José Olympio , 1985.

Haeckel, Ernst. *História da criação natural ou doutrina científica da evolução*. Porto: Leloo & Irmão , 1961.

Machado de Assis, Joaquim Maria. "Notícia da atual literatura brasileira: Instinto de nacionalidade." Machado de Assis, *Obra Completa*, vol. 3. Rio de Janeiro: Editora Nova Aguilar, 1994. Editoração eletrônica: Gabriela Balieiro Moreira Nilza Pieritz. Acessed in June of 2007 (http://www.cce.ufsc.br/~nupill/literatura/BT4522147.html#[4]%20NOTÍ CIA%20DA%20ATUAL%20LITERATURA%20BRASILEIRA).

McCann, T.S. , M.A. Fedakz, and J. Harwood. "Parental Investment in Southern Elephant Seals, *Mirounga Leonina*." *Behavioral Ecology and Sociobiology* 25 (1989):81-87.

Pereira, Lúcia Miguel. *Prosa de ficção (de 1870 a 1920): história da literatura brasileira*. Belo Horizonte: Itatiaia; São Paulo: USP, 1988.

Ridley, Matt. *Nature Via Nurture: Genes, Experience and What Makes Us Human*. New York: Harper Colins, 2003.

Sevcenko, Nicolau. *Literatura como missão: tensões sociais e criação cultural na Primeira República*. São Paulo: Cia das Letras, 2003.

Shwarz, Roberto. *Um mestre na periferia do capitalismo: Machado de Assis*. São Paulo: Livraria Duas Cidades, 1990.

Skidmore, Thomas. *Preto no branco: raça e nacionalidade no pensamento brasileiro*. Rio de Janeiro: Paz e Terras, 1976.

Sodré, Nelson Werneck. *História da literatura brasileira: aspectos econômicos*. Rio de Janeiro: Civilização Brasileira, 1964. 381-402.

Trivers, Robert. *Natural Selection and Social Theory: Selected Papers of Robert Trivers.* Oxford: Oxford UP, 2002.

Vincent, Amanda C.J.;, and Rosie Woodroffe. "Mothers Little Helpers: Patterns of Male Care in Mammals." *Trends in Ecology and Evolution* 9 (1994): 294-297.

This work is part of the project "Two physicians, two writers: the uses of Darwinism in Brazil (1875-1900)."

CHAPTER 9

José Martí and Evolution: An Analysis on Nation and Race

In 1891 the Cuban intellectual José Martí made public his ideas about the identity of post-colonial Spanish America in a text that appeared in Mexico City under the title of "Nuestra América" ["Our America"]. This text has had many interpretations, some of them dealing with the beginning of a new way to understand the political realities of the region. José David Saldívar, for example, mentions that in Foucault's sense, "Martí stands between two ways of thinking: the last representative of a nineteenth-century romantic idealism and the first forerunner of a Latin American socialist ideology of continental solidarity" (*The Dialectics* 7). But, among all the possible interpretations that a text of this magnitude can have, there is space to think about it also as an attempt to develop a new understanding of identity after the introduction of Darwinian ideas in the region.

The gaze that was popularized after the publication of Origin of Species in 1859 was focused on the observation of variation, and its importance in establishing evolution. Unity was located in the past, at the moment of origin at which all species converged. The future, by contrast, entailed divergence, struggle, conflict, and the impossibility of permanence. Extinction was the other side of speciation. Life was ultimately the result of death. This new scientific perspective conflicted with the needs of those who were in the process of forming or organizing new nations. So, all sorts of confusing and contradictory ideas emerged to try to resolve this conflict between permanence and change, spirit and matter. Eventually the racialization of nations, the idea that the nation rested on racial development toward a final and distinct type, would become the center of evolutionary political thinking.

José Martí lived through the years of this transformation, but the intersection between his ideas and Darwin's has not received sufficient attention. As is often the case in research concerning Latin America, the study of philosophical and political ideas does not take into account scientific thinking; a curious fact, since the process of nation-making during the second half of the nineteenth century saw philosophy and politics as subordinated to science. No intellectual who wrote during this time was immune to scientific thought, particularly after the publication of Darwin's work. Martí was no exception. Rafael Catalá has mentioned the presence of an evolutionary dimension in the poem "Yugo y estrella." In his intertextual analysis of *La Edad de Oro*, Luis A. Jiménez also notices how the description of animal behavior followed Darwinian motifs. Jorge Camacho's work analyzes at how many levels Martí's writings reveal an evolutionary concern that is not that different from other members of his generation. According to him, Martí formed his racial vision "inside the scientific views of nineteenth century evolutionism" ("Caminos"). Following this line of inquiry, in this essay I will explore how Darwinian ideas intersected with the work of Martí, particularly in his understanding of the relationship between race and nation. I would like to show how his response to evolutionary thinking can be divided in two distinct ways. First, we find a spiritual conception of nation that expresses a strong rejection of evolutionary concerns about race; the second, which is represented by "Our America," invokes a language much closer to that of evolutionism, and I propose that this can be read as a neo-Lamarckian text. This transition responded not only to the political changes in Cuba; it was also articulated within the scientific debates that Martí closely followed.

The World that Martí Inherited: Darwinism, Race and Imperialism

The Darwinian revolution transformed the way in which nature and the very place of human beings within it were conceived. This radical transformation of science impacted political thought beginning in 1859, the year in which *Origin of species* was published. Since science had been associated with the process of building a civilized society since before Darwin, his views triggered important debates as to how they could be used to understand *human* evolution. The problem was that the English naturalist did not systematize his ideas in the accustomed manner. For example, he was interested in universal causes to explain variation in nature, but he did not attempted to explain how this variation would lead to a better future. Natural selection did not serve the great cause of progress, but the more humble cause of survival. Darwin insisted on the importance of heredity, but did not provide a mechanism for understanding this process. His implicit attacks on previously held views on unity and harmony in nature, and on the notion of universal progress, were not complemented with examples that made room for the particular adaptations to political thought.

Darwin-inspired literature contradicted the emphasis on harmony and cooperation among individuals with its emphasis on struggle and competition in nature. More importantly, since this theory was concerned with variation, and variation among humans was understood mostly in racial terms, race was at the center of human evolutionary preoccupations. In this way, nations represented peoples who were separated by the reality of their racial past, and in order to prosper these groups had to struggle against each other. So, as Charles Hale has mentioned, the Latin American preoccupation with race "was further strengthened by the Darwinist aspect of Spencer's thought and that of other influential evolutionary theorists such as Ludwig Glumplowicz and Ernst Haeckel" (163).

In racial terms Darwin himself would introduce a new perspective (in *The Descent of Man* and related texts), arguing that racial differences were not the result of natural selection, but of cultural preferences. But this explanation was not consistent with the political culture of the time, and was seen as inconsistent with the evolutionary prevalence of natural selection. The contradictions that the new evolutionism introduced explain the importance of Herbert Spencer, who in the development of his synthetic philosophy provided a totalizing system to clarify previous ideas of progress while incorporating evolutionary views. The same would happen with the work of Ernst Haeckel, who eased racial uncertainties with his evolutionary validation of racial categories. In the words of Mario Di Gregorio, in Haeckel's view, "the language step was followed by a "nation-making" one, so that language, the idea of nation, the concept of species and races, were all united," providing a synthesis that was relevant for those who were interested in translating biological evolution into political results (255). The assimilation of race and nation meant that the emergence of a national type was needed in order to reorganize the country. Purity was privileged over hybridity, a category that was considered evolutionarily suspect.

The early emphasis on natural selection, racial competition, and the survival of the fittest shocked the Latin American thinkers who supported European science, since it represented a manifest contradiction with the ideals of the Independence Movement and the first post-colonial discussions of nation building. The 'civilized' world seemed to them split between ideas championed in the work of Humboldt, and the news ones propounded by Darwin. The chaotic post-Darwinian world, in which basic working elements of evolutionary theory, such as heredity, remained unclear, created a fragmented sense of reality dominated by uncertainty. We see the same among writers in Europe or the United States, and for similar reasons. Those who sought to deal with the rupture induced by Darwinian ideas had to face the fact that they were not part of a grand narrative in which past, present and future formed a universal sequence. As we will see in the case of Martí, the main problem that becomes evident by the 1880's is the lack of the philosophical synthesis needed for building the nation.

Cuban writer Carlos de Pedroso, an anti-Darwinian, affirmed in 1885, that as a philosopher "Darwin was of the smallest size you can conceive;" his importance was based "purely on his genius for observation and, in my opinion, without any synthetic depth." More importantly, the author of *Origin* was unable to comprehend the meaning of the events he described and, even more frequently, "he did not see the important consequences his observations lead to" (47). Contrary to Herbert Spencer, Darwin was not concerned with creating a totalizing narrative that explained how the progress of the nation would be achieved in accordance with Nature's plan. De Pedroso also points out that another problem with the modern idea of creation, as opposed to a traditional religious notion, consisted in its reliance on an analogy to the rationality of human choice, serving in this sense to justify the very anthropocentric visions that it supposedly opposed. This fair and correct criticism targeted a naturalization of events by recourse to a logic that was not that of nature, but of human power.

> The idea that the concept of a plan and the notion of continuity in a regular and successive manner are contradictory is a capital error that dominates and directs the Modern School [evolutionism]; an error born from the excess of its own anthropomorphism that this same school rejects and accuses its opponents of indulging. In effect, many [evolutionists] often write as if design, intention, and purpose applied to Nature were necessarily of the same kind and character applied to the operations of the human mind. [...] The idea of "plan" or "design" [...] is simply the word used to express a kind of manifestation of order that expresses a group of regularized and subordinate facts (125).

In this anthropomorphic way of thinking, the logic of natural selection applied to nations promoted analogies that implied that in the most advanced among them natural selection led to progress, while those who could not compete were being culled. In this way France's defeat at the hands of Prussia in 1870 seemed to indicate that the former's loss was due to a process of degeneration caused by its weak racial constitution. The emergence of the United States as a world power also seemed to show, according to this view, that nature was favoring the north European Anglo-Saxons as the group that was most fit to develop civilization. This country's superiority was supposedly not based only on its industrial progress, but on its possession of a better population selected by nature to dominate and control people and resources. Evolutionary ideas and politics were in this way connected to an ideology that emphasized the special role of the United States in the world. David Starr Jordan, a prominent voice in North American intellectual circles, provides a good example of this connection between science and politics. He wrote in 1909, while president of Stanford University, that just fifty years ago Darwin had written *The Origin of Species*, an event that predated by one year the triumph of Lincoln as president of the country.

> ...and his unique personality [Lincoln's] began to stamp itself on the nation of democracy, and on the world, which is in turn a democracy in proportion as its people are competent units, fit to take care of themselves.
> "The Origin of Species" has changed the entire face of the science of life and its philosophy, which must ever be the outgrowth of science. It gave to scientific men, with its demonstration of the origin of living forms, the first clear explanation of the phenomena of orderly change. [...] Before Darwin, biologists had catalogued, compared, and classified; but to their questions of "Why?" no answer could be given. (286).

As Jordan makes clear, the identity of his country emerged at the same time that the new evolutionism was being promulgated, and Darwin's ideas and the United States were linked and destined to impose themselves over other nation's designs. It was the creation of a new order that used science to explain the logic of political domination. But even in North America, not everyone praised the conjunction of Darwinism and Imperialism. In 1905 Columbia professor James Hyslop wrote an essay entitled "Why are we Imperialistic?" In it he affirmed that the "primary influence in the modern movement may be dated in 1859," the year *Origin of Species* appeared. According to Hyslop, the publication of this work, and the debate that followed, led in the United States to abandonment of old ideals related to non-economic preoccupations (2). The country had “drifted away from the religious ideal of the past to that of the economical, and Evolution has been the most effective agency in relaxing the allegiance of men to the older ideal” (2-3). The only way to return to a more balanced society, one that incorporated the ideas of the past on which the nation had been founded, was to return to a more spiritual foundation. For Hyslop nothing "but the recovery of a spiritual ideal will redeem us from the rule of imperialism" (15). Jordan and Hyslop saw the role of science in politics differently, but what is clear, as Michael Hunt explains, is that “Darwinian notions served to reinforce preexisting ideas of Anglo-Saxon superiority.” According to the latter, “lesser races, awed and grateful, could follow the lead of the Anglo-Saxon- or drop to the bottom of the heap to meet their fate, ultimate extinction” (79).

In terms of evolutionary thinking the 1870’s were characterized by the multiplicity of contradictory views that dominated the debate on natural selection. There was a sociological perspective, introduced by Herbert Spencer, along with biological and anthropological perspectives. From the point of view of nation-building, this is a chaotic picture that created great uncertainty. The incomplete state of knowledge about the mechanisms regulating heredity originated all sorts of intellectual debates, particularly among naturalists and anthropologists. As anthropological views were connected with the understanding the logic of evolution and extinction, psychology provided “another dimension to nineteenth-century race consciousness,” particularly through the works of Hippolyte Taine (Hale 163). French social psychology was instrumental in restoring a notion of historicity that had been neglected by Darwin’s account of evolution. Since for the latter time was not a universal category–different species might each evolve at a different pace–universal progress was not applicable. But in Taine’s view, history is again a force that implies uniform results over a given span time. According to Aníbal González the source of Taine’s attraction for Spanish American critics is that he “provided his readers with a seemingly systematic critical vocabulary with which to approach more familiar and contemporary texts." In addition, his "implicit Hegelianism was probably another source of interest, given the Spanish writers' constant preoccupation with their national history and its place in a broader historical scheme" ("Literary Criticism" 440). In the case of Cuba, Taine's work was received in the context of the evolutionary debates. In 1897, after José Varona, editor of ***Revista Cubana***, left for New York, Juan Bautista Casas praised his journal as the source of wisdom for "philosophers, sociologists, and independent scientists." According to Casas the intellectuals who published in it prided themselves on belonging to the schools of "Lombroso, Darwin, Taine, Spencer, Stuard-Mill [sic] and Virchow" (92).

Taine's writings were extremely important since they introduced the triad of race, milieu, and moment common to most intellectuals interested in politics who published in the 1880's. José Martí, for example, would use Taine's ideas in his 1887 essay "Walt Whitman" (Saldívar 441). Modernist poets were also familiar with his views. More importantly for the purpose of this essay, Taine's biological basis for his definitions of these three terms relied on Darwin's ***Origin***. But his use of Darwin's theory attempted to modify aspects of it by incorporating into it a totalizing system that included a political vision well beyond Darwin's. Ross Shideler mentions that in notions of authority and language Taine followed a path similar to Auguste Comte, which means that he was another thinker interested in creating a comprehensive system that organized all aspects of society (34).

French influence continued to be critical during the 1890's, particularly in the re-emergence of Lamarckian evolutionary ideas through the interest in the inheritance of acquired traits. Jan Sapp places the beginning of neo-Lamarckianism, which I will discuss later in the context of Martí's work, in France in the 1880's "when Darwinian theory, which had previously included the notion of the inheritance of acquired characteristics, was beginning to be identified solely with natural selection" (124). This reaction was in part also a rejection of the logic that justified the supremacy of the most powerful by recourse to natural selection. In fact the return to the work of Lamarck in France was connected with the defeat of 1870, which had "brought a patriotic tendency of French biologists to encourage support of their own theories rather than accept or develop those of other countries, especially Germany" (Sapp 124).

As the centrality of Darwin's original exposition declined, the evolutionary debate was taken over by those who saw evolution as related with natural selection and hard inheritance, centered around the work of Auguste Weissmann; and those who supported neo-Lamarckian ideas. Because of this, by the 1890's evolutionary thought became more helpful to those in Latin America who wanted to establish a relationship between science and nation-building. This was the world in which Martí emerged as an intellectual interested in overcoming the legacy of Spanish colonialism. As I will try to show, some of the events just mentioned here would have an important impact on his intellectual development.

Cuba, Martí and Darwinism in the 1880's

By the 1880's evolutionary thinking dominated the intellectual circles of Spanish America. But, as we have seen, the state of evolutionary thought was in flux. Ideas coming form Charles Darwin, Herbert Spencer, and Ernst Haeckel, among others, all coexisted. La Habana could be identified as arguably one of the most important centers of scientific discussion in Spanish America. It had an extremely close relationship with European institutions of knowledge and new ideas reached this city extremely fast. In 1886 the Cuban writer Rafael María Merchan mentions in his study of the cultural life in the Cuban capital that in this country "the models were the French, English and German poets. The same with philosophy, and it is certain that if tomorrow a new doctrine overpowers Darwin's ideas and similar schools of thought, the institutes of education and cultural gatherings in La Habana will adopt it…" (711). In fact the enthusiasm for the new science was such that in a compilation of poems by Juan Martínez Villegas, published in 1885, we find one entitled "Are men descendants of monkeys?" The editor of the book explains that Villegas had written this strange poem a few years earlier, "when in Havana it became fashionable to talk about the grossly materialist ideas of the famous Darwin" (315).

In Cuba the discussion of Darwin's ideas, and their political and philosophical consequences, began in the 1860's. Francisco de Frías y Jacott, Count of Pozos Dulces, wrote an article on *Origin* in 1868, around the time of his famous speech at the Academia de Ciencias Médicas, Físicas y Naturales de La Habana. This analysis of Darwin's ideas, which he had read in English, touches on all the problems that would be debated in the years to come. First, we find a criticism of a method that relies mostly on analogies in articulating the importance of natural selection. Second, Frías y Jacott understood the importance of Darwin's narrative on variation and its plots, as a reason why the argument was so persuasive to some, even when it presented so few proofs, particularly regarding the workings of heredity. In this way, *Origin*'s hypotheses were "fantastic dreams," and the result of Darwin's "fecund imagination" ("Sobre el Origen" 217-218). Liberating science from the traditional rigor of the laboratory and the control of the scientists, Darwin made it possible for anyone with an interest in observing nature to deduce its workings from the analogies between natural and artificial selection. This creation of possible plots, related with mating and lines of descent, increased the interest in this theory and its popularization.

In 1874 Francisco de Frías changed his assessment of natural selection and variation and accepted Darwin's ideas. In a letter written in 1874 he mentions a previous article in which he had defended the interests in educated women on two separate grounds. First, "they are the first and best educators during infancy;" second, they are instrumental in "the formation of races more intellectual and morally prepared than the ones obtained by the current system that concentrates in the education of boys. If Darwin's and his followers' theory is right –and I believe it is- the intellectual and moral attitudes are transmitted from generation to generation in the same way in which physical aptitudes are passed." This last point was linked to the need to correct a process of heredity that was imperfect or incomplete "when only one of the parents, the father, carries the aptitudes inherited or acquired by education" (540).

De Frías's change of mind is explicable by recourse to two events that illustrate the importance of Darwinism in Cuba, and in general in all Latin America. In 1871 Darwin's emphasis on soft inheritance and sexual selection in *Descent of Man* opened the possibility of a more benign idea of heredity than that represented by natural selection. his more open acceptance of the inheritance of acquired traits allowed for the introduction of Lamarckian ideas, also defended by Herbert Spencer, in a way that tended to promote the importance of culture and education in changing the character of a given population. But at the same time that soft-inheritance was becoming more prominent in Darwin's work, in some evolutionary discussions there was also a tendency, as a result of political changes, to privilege natural selection, and a harder view on inheritance. As I have mentioned, the defeat of France in 1870 was perceived as evidence that natural selection was favoring the Anglo-Saxon over the Latin race. Together with the emergence of United States as a powerful nation, this event seemed to confirm that competition among nations followed racial lines. So, even when Darwin gave soft-inheritance a larger role in evolution, the emphasis of the 1870s was on the supremacy of natural selection and its role in building the most powerful nations.

In the midst of the various disputes and corrections of the post-Darwinian debate, there was an important series of lectures in which José Martí participated actively. In 1879 he was one of the organizers of a series of lectures on "Idealism and Realism in Art" at the Guanabacoa Lyceum, which continued with a debate about the origin of man. At this point Martí accepted Darwinian observations about nature, but rejected the use of analogies that reverted to a vision of society dominated by materialistic concerns. Pedro Pruna sees in his preference for ethics and idealist doctrines Martí's need to "oppose the utilitarianism of the North, and the emerging pragmatism that he connected, correctly, to utilitarian positivism" ("La Recepción" 379).

Participants in this round of talks included Enrique Varona, a defender of Darwin and Spencer; and representatives of positivism, such as Francisco Arango and the secretary of the Academy of Medicine, Antonio Mestre. The ensuing debate indicated the importance of the new evolutionary ideas and the vitality of the response with which the introduction of Darwin's work had been met in the island. This meeting sparked a controversy among Spanish colonialist newspapers, "and also by a group of seventy distinguished ladies, who were outraged by the fact that the thesis on the animal origin of the human species had been expounded before an audience that included women and children" (Pruna, "Biological Evolutionism" 56). Regardless of this opposition by 1880's Darwinism was part of the teaching at the Real Academia de Ciencias Médicas, Físicas y Naturales de La Habana. Other institutions would eventually follow. The Anthropological society would organize an annual conference to honor Darwin.

Reflections on the nature of Darwinian materialism irritated many intellectuals who were divided on this issue. Martí warned, in no uncertain terms, about the consequences that evolutionary thinking would have for Spanish American nations. For others, such as Antonio Mestre, there remained the possibility of reconciling the materialism of Darwiniist ideas with more spiritualist views closer to the Cuban intellectual tradition. In his speech at the Guanabacoa meeting he argued against those who criticized materialism, asserting that they did not understand that one "could be transformist without denying human duality" (527), a position closer to French evolutionism. He pointed to the revolutionary way in which Darwin's theory had changed the meaning of nature and evolution. Harmony and beauty coexisted with "the combat for existence;" beneath the appearance of harmony there was "a never-ending struggle" triggered by forces of "destruction." So long as nature seemed to be breathing "peace and serenity," people failed to noticed that all happiness "rested on a vast and eternal cemetery" (427).

This relationship between Darwinian ideas and death would be a common theme all over Spanish America. Since the existence of evolution was related with the complementary loss entailed by extinction, in order to create a nation along evolutionary lines some lineages must perish, a process that announced struggle and violence along racial lines. Since evolution implied both continuity and discontinuity, it was related with the problem of conditions under which a nation could persist, which explains why, according to Pedro Pruna, one aspect of the Cuban debate on Darwin's ideas "is obviously associated with politics." The Autonomist Party, for example, adopted the slogan "all through evolution, nothing through revolution" by the end of the nineteenth century in an attempt to curtail the push for independence ("Biological Evolutionism" 58). Darwinians who favored Independence, such as Manuel Sanguily, accused the autonomists of twisting Darwin's thought. But since the English naturalist's analogies permitted adaptation to multiple contexts it was possible at the time to think that progress was not due to human action, but to natural (racial) evolution over time. As Pruna shows, Sanguily, in order to contradict the Autonomists, wrote that Darwin "was in favor of struggle," and it was undeniable to him that "revolution is a necessary and fatal form of evolution" ("Biological Evolutionism" 58). Contradicting those who favored continued colonialism while the nation was formed through gradual biological evolution, Sanguily reminds his readers that without struggle there *was* no evolution, and that the revolution was the political expression of this type of conflict. Cuban intellectuals "were overwhelming in favor of the new, Darwinian (*sensu lato*) approach and this allowed for more daring enterprises, such as the open explanation of Darwinian (*sensu stricto*) in university courses" ("Biological Evolutionism" 58).

The fact that it was mostly the Spanish colonialists who opposed the new ideas made this debate particularly intense in Cuba. In 1907, on the occasion of the dedication of a statue of Lamarck in France, Carlos de la Torre remembered how thirty years earlier the "immortal work" of Darwin had been the object of "passionate controversies," and how Francisco Adolfo Sauvalle had delivered a famous address on "La continuidad de la Naturaleza y la descendencia del hombre" at a session of the Academy of Medical Sciences in 1875 (330). The *Revista de Cuba* had an important role in promoting the discussion of evolutionary ideas. In 1880, for example, it published a lecture of Huxley's given only a few months earlier. The following article reproduces a talk given at the Anthropological society about the ideas of Wilhelm Wundt. In it José Francisco Arango argues against Wundt's work. Arango violently rejects any idea that meant a return to a type of metaphysics that he identifies with the past. He denounces any type of psychological spiritualism as a return to the "absurd superstitions of the Middle Ages, at a time when the scientific spirit is creating such monuments as cellular theory as propounded by Virchow, transformationism as propounded by Darwin, and evolutionism as propounded by Spencer," not to mention "positivism according to A. Comte and Littré" (33). Being a positivist, this author rejects the German idealist revival that started to emerge from the study of psychology as an ideology of the past.

Since in Cuba the ideas derived from Darwin were so important for the future of the country Martí followed closely the debates on this issue. But he also knew about evolutionism through his stay in the United States, and about the importance that Darwin-inspired ideas had in the building of imperialism. In one of the articles published in *La América* in 1883, for example, Martí mentions how at a New York school a speaker had called for a replacement "of Homer by Haeckel; Greek by German; metaphysical by physical arts." He notes that this demand is "a password [contraseña] of the times, present in every newspaper and relevant magazine." Martí calls for the teaching of evolutionary science in order to understand nature. Scientific instruction must become like the "sap of trees, flowing from the roots to the crown of the system of public education" (OC VIII,

278). Darwin's influence on Martí is undeniable. He praised the naturalist's vision, his acuity of observation, and the new discoveries that were published monthly because of his theory.

In an article entitled "Un Mastodonte," [A Mastodon] also published in La *América* in 1883, he praises the Darwinian renewal that now allowed paleontologists to "publish wise comments." He continues by expressing the pleasure he takes about the work of these men, and his sense that this was the beginning of "an epoch of true revelation" (OC VIII, 410). But while he supports paleontological work that reveals the context in which species had become extinct, he remains concerned about how the same ideas might be applied to human beings, particularly those who were still alive. In a letter written in New York and published in Mexico's *El Partido Liberal* and Buenos Aires' *La Nación* in 1887, Martí reports on a scientific gathering in the United States at which Darwin's theory had been discussed. He concludes that the speeches given during this meeting were incomplete, since some scientists forgot that "every theory about the body has to be proven by one related to the spirit;" while others, "self-centered and arrogant," overlooked the fact that the relationship between body and spirit was no different than that between "sublime music and the feeling that it expresses," a fact that explains why music "will never go extinct." He concludes that "everything refines itself, purifies itself and grows" (OC XI 278)

Martí's rejection of Darwinian ideas as applied to humans was related with the emphasis that the analogies of natural selection placed on the naturalization of conflict and separation. If the process of civilization was constituted by competition and the struggle to survive there was no possibility of imagining a political change that would lead to a more just society, one that promoted the common interest above individual needs. In this sense, Martí's view could not be separated from the pre-Darwinian ideas that associated the culture of progress with the gradual perfection of society, and the ultimate unity of all human beings. Post-Darwinian conceptions of race were an obstacle to envisioning the future. In the particular case of Cuba, and in general in the post-colonial Spanish republics, they meant that the organization of the country depended on natural processes and not on the political will of individuals. There was a need, for Martí, to transcend the materialism of race to find a way in which colonial bodies could be transcended in the emergence of the republic.

Martí's interest in transcendental thought is thus related with his need to articulate a new way to understand the Spanish colonial system and post-colonialism. Part of this reflection originated in what was happening in Spain itself, since the most progresive Spanish Liberals "soon realized that in order to combat the conservative and traditional elements, Spanish liberalism had to have an idealistic basis" (Stoetzer 90). As it is well-known, while studying in Spain Martí came in contact with the writings of Julián Sanz del Río, the translator and promoter of Krause's work, which had never been very popular among German intellectuals, in part because he was a contemporary of Hegel's. But for the Spaniards it provided an important tool for opposing the materialist currents that dominated the culture of the time. This is what explains the success of Krause's ideas in Spain, "since the new Liberal breed saw in a Spanish adaptation of Krausean philosophy- krausismo- an effective way to fight both the conservatives on one end of the political spectrum and the Positivists on the other" (Stoetzer 91). In a similar way, his philosophical approach was of interest for Martí since it introduced a "religion of humanity with a new vision of man as the synthesis of the universe" (Jorrin 91). To the ugliness and confrontation that resulted from the Darwinian understanding of nature, this vision maintains the importance of harmony and beauty.

Martí's views on nature strongly rejected material reductionism. In all the works of nature one might observe "the coexistence, the interdependence, the inter-relation of matter and spirit!" (OC XXIII, 317). The relationship between these two concepts orginates, as Susana Rotker mentions, a work that is "a field of contradictions" (88). But it is important to remember that some of these contradictions were not of Martí's making. Thinkers in the United States and Europe labored under the same confusion as to how to reconnect modern science with ideas expressed before and after Darwin. As Martí noticed, Darwinism had eliminated old philosophical traditions without replacing them. He wanted "spirit to win over matter," but he was well aware that the possibility of this triumph was compromised by the dominant materialism of his time (Rotker 88).

Martí's emphasis on the spirit in his account of race was related, among others sources, to French authors influenced by Victor Coussin. Ernest Renan provided a bridge to this tradition. It is interesting to note that Renan's work attempted to neutralize the impact of Darwinian theory, but this did not imply an abandonment of evolutionary ideas. In Alberto Spektorowski's view, although Renan "believed in the future intermixing of cultures and in the triumph of humanity and universal reason, the basic association of environment and cultural development as a universal condition could hardly be discarded" (123). In promoting a universalism that was questioned after Prussia's defeat of France in 1870, Renan's ideas were in part an answer to Germany's racial policies, which he warned would lead to zoological wars of extermination "analogous to those which various species of rodents or carnivores participate in to survive" (qtd. in Rabinow 127).

Renan's influence was also helpful in solidifying faith in human duality. But, more importantly for Martí, he criticized the association between nation and race, and natural selection with evolution. According to the Renan, the substitution of an anthropological principle by a national one was a great error that "would destroy European civilization. The primordial right of races is as narrow and as perilous for genuine progress as the national principle is just and legitimate" (13). Interestly, as Kathryn A. Manzo explains, Renan argued that "nations should be constituted by affective sentiment (which can change) and not be based on whatever is presumed to remain fixed" (58). Race, then, could not be the basis of any nation. Blaming the emergence of antrophology as a discipline

Renan questioned the use of "the term race as synonymous with family, to indicate a blood relationship" (58). He intended to detach race from nation at a time in which even those "who recognized the "mixed blood" of their own populations equated the formation of new races (from racial mixing) with the formation of new nations" (Manzo 59).

Of all the writings of Renan, of particular importance to Martí was "What is the nation?" originally given as a lecture at the Sorbonne in 1882. This piece contains strong denounciation of race as the primary category in the process of nation building. As Ernest Gellner suggests, the main purpose of this talk was to challenge the deterministic understanding of nationhood that prevailed at the time. According to this view, nations were not the product of any deterministic process; they are not "dictated by language, geography, race, religion, or anything else." He clearly rejects "the spectacle of nineteenth-century ethnographers as advance guards of national claims and expansion. Nations are made of human will" (8). In a letter of April of 1882, Martí mentions the Sorbonne lecture, that had taken place only a few weeks before. Renan had, he writes, shown up as laughable the view that men must be guided "by what is called the racial spirit [espíritu de la raza] when the human spirit is so magnificent and splendid that in it all men have communion at the same time," a fact denied by those interesting in promoting hate and conflict (OC XV, 449). Quoting directly from Renan, and providing his own translation, Martí directs the reader to a strong criticism of the post-Darwinian world.

> Human history–says Renan--is not a chapter of zoology. Man is a rational and moral being. Free will stands above the ruinous suggestions of the racial spirit. A nation is a soul, a spiritual principle, made in the past, with life in the present, and all the great meetings of men with healthy minds and generous hearts can create the moral conciousness that makes the nation (449-450).

Closing on this concept Martí anticipates a time in which nations will be at peace, and when all the men on earth, given to love each other, would feel "in their robust chests" the benefits that come "from virile human love!" (450). Interestingly, belying the universality that imbues Martí's reading of Renan, he makes it clear that the love that was needed was that of *men*. Perhaps a concern with feminization as a sign of evolutionary weakness lay behind his emphasis on "virile" love, but in this sense he follows the mysoginistic logic so prevalent in the intellectuals of the late nineteenth century. But despite the emphasis on virility, the men he describes in the following section seem to be in a process of decomposition. On the one hand, the men who announced the new era were "poets, the knights of peace," who as "unquiet spirits," [espíritus inquietos] "whipped and consumed in flames the bodies that constrained them;" but in the next sentence, the physical forms of these spirits turn out to be men "of pale color and deep gaze;" "sick heroes" announcing "a magnificent time, a distant epoch" (450). It seems that even in the act of mortifying their bodies, the weight of their racial inferiority, these men cannot act in the present. Like specters, their promises speak only to a new time in some remote future.

I think that this sense of paralysis in concrete present action is related with racial ideas derived from evolutionary thought, and Marti's own investment in them. On the one hand he sees the solution to the problems of the present in eliminating the notion of race, but on the other, at a certain level he believes that some individuals are better than others. This will be, by the way, typical among Spanish American intellectuals at the time, who criticized any ideas of inferiority that included them, but also expressed a belief in the inferiority of certain groups, particularly when talking about culture. I agree with Lillian Guerra that Martí's idea of whiteness "was a matter of culture, not color. Anyone might acquire that culture, if only they were willing to try." He carefully confirmed "the continuing superiority of educated white Cubans and invited black Cubans to feel included in a multilayered vision of nation that could accommodate and nurture the virtues of both groups" (34).

So, Martí's dispute was not with the inferiority of certain groups, but with an explanation of this inferiority based on racial determinism, something permanent and fixed. It can be argued that what he is doing is denying that a culture belong to a race, and affirming that there is no Western thought, but rather a universal thought that belongs to all men. But then what place is there for difference? If he allows for different types of bodies to live together, the same surely does not hold for the coexistence of cultures. Lillian Guerra provides examples that show that the elimination of African traditions by Afro-Cubans is viewed with great relief by Martí (34). His concern for universal progress determines many of his cultural views. But, interestingly, this obsession with progress does not lead him to accept Herbert Spencer's ideas, a telling fact considering that the latter was the most important figure related with the restoration of a faith in eternal progress after the Darwinian revolution.

Unlike many of the other members of the intellectual elites, Martí sees that Spencer's identification of progress with struggle may prove very destructive for Spanish American nations already consumed by division and violence. In his view a restoration of a universal perspective, a spiritual understanding of the basis of association was the only way to unify the nation. In a commentary on Herbert Spencer, written in the 1880's, Martí concludes that the latter's ideas promoted a view that would only bring conflict and destruction to human societies. In his exclusive attention to human harmonies the English sociologist had lost track of the "more vast and fundamental [harmonies] of nature." Moreover, to encourage different peoples to occupy themselves exclusively with their earthy goals and future was to "corrupt them with the best intention of curing them." Those populations who did not believe "in perpetuation and universal meaning," in the "glorious ascent of human life," run a risk of desintegration, like "a piece of bread eaten by mice" (OC XV, 388). The clear target here is Spencer's view of the struggle for life. Marti clarifies that "human life" needs to ascend, thus opposing those evolutionists who talk about the ascension of the favored races over the less favored. In his spiritual view, nature does not operate through selection and culling, but by creating harmony in the environment.

Our America as a Response to Darwinian Concerns

In 1891, Martí published in *El Partido Liberal* of Mexico an essay entitled "Nuestra America." This work was an expression of the journalism of the time, which "was placed in the service of nation-building," as Aníbal González reminds us (*Journalism* 15). In this work we find a new way to look at the continent, one that allowed for the simultaneous existence of difference and sameness, change and permanence. But, importantly for the purpose of this essay, the use of evolutionary ideas seems here much more important than in other texts of Martí. Race is dissociated from natural selection and materialism, which allows him to see the process of struggle in nature as part of the future evolution of Spanish America. Acording to Julio Ramos, in the teleology of "Our America" "history is not seen as the harmonious becoming of a future perfectability, but rather, as a process of continuous battles," an attitude that I relate to cumulative impact of scientific evolutionism as evidenced in this text (254).

Jorge Camacho has observed that the texts written in the period between 1886 and 1890 offer a different racial rethoric. According to him this change is related to preparation for the war of Independence and its strategic needs ("Signo" 77). But alongside these political needs important changes were taking place in the evolutionary rhetoric of the time, making room for the emergence of a movement that would place the burden of evolutionary change on the influence of the environment and human will. I do not think that racial thinking was the same in Latin America after the popularization of neo-Lamarckianism; in the same way, Martí's writing about race was certainly not the same by the 1890's.

A central protagonist of the future change in "Our America" is the "natural man" [hombre natural], who obeys and rewards "superior intelligence" as long as it does not exploit "his submission" to hurt him, or offend him through his exclusion, an action "that the natural man does not forgive" and tries to avenge. Natural man is good, and willing to accept the guidance of the cultured as long as they do not disrespect him with ignorance of his reality. In this sense there is in this text an interesting intersection of perspectives. The natural man is ignorant and must be educated, but those who want to educate him need to understand his reality. There has to be an acceptance of the conditions of existance of those who live closer to nature than to reason; but those closer to nature must to accept the need to change, to improve themselves. In this sense, this was a conflict of visibilities. Neither of the opposing parties could deny the existence of the other.

If in the discussion of Renan's work the nation is elusive and located sometime in the future, in "Our America" it has been born and is located in the present. The post-colonial reality is related not to disease, but to life and possibility. Even when there is a duality that expresses differences, and there is a reference to struggle, these are resolved in a coordinated dialogue. Culture has the same importance as nature in this text, and the success of one depends on the agreement of the other. For Martí, the battle was not between civilization and barbarism, but between "the false erudition and nature" (*OC* VI, 17). There are also other important differences. The first is related to the role of nature, and the replacement of the matter/spirit duality with that of nature/culture. In the interaction of the latter, there is an evolutionary logic in the sense that for something to be fruitful it must come from an adaptation to the environment. A culture that is not naturalized, which means adapted to nature as part of its manifestation, cannot continue. This explains why for Martí the natural man triumphs over civilized ideas. Matter/nature is at the center of the analysis of adaptation. In fact it is quite surprising, in view of Martí's earlier ideas, that in discussing the future of the region a mention of "spirit" is absent.

The other difference between "Nuestra America" and Martí's views of the 1880s is his newfound optimism. In fact, he talks about the future formation of the "real man." No longer merely a vision, the product of the imagination, he is made by the "serene harmony of nature," on the "continent of light," and through the influence of recent European work, no longer the result of "approximation and uncertainty," but of reality. This last remark is rather important since it expresses a recognition on Martí's part of a new way of understanding nature among European thinkers. The previous era of "approximation and uncertainty" probably represents an earlier phase in evolutionary debate, when discussion focussed on the role of natural selection.

The new era to which Martí alludes is, I believe, one shaped by the rising importance of neo-Lamarckian ideas in Europe, and their interpretation of Darwin in accordance with soft-inheritance and environmental influence. As I have noted, discussions of Darwin himself are not the same in the 1890s as they had been in the 1880s. By now Darwinians were split on the rol of natural selection. Those who opposed the supremacy of this mechanism were called neo-Lamarckians and believed that variations were "based upon environmental differences and the force of will and reason in equal amounts" (Aitken 41). Martí knew about the different views on evolutionism attributed to Darwin and Lamarck. In an article for the *Opinion Nacional*, written in 1882, he mentions that in a speech a naturalist had referred to the "true father of evolutionary theory [as] the famous French naturalist Lamarck," while Darwin and others were just following in his footsteps (*OC* XXIII, 287). This interpretation was even more predominant in the 1890s, when an important scientific current sought to blend the ideas of these two figures.

It is doubtful that Martí would have been unaware of these recent developments. In Cuba, Lamarck had been known since the 1820s. In his work on botany, published in 1824, Ramón de la Sagra mentions the French naturalist's work as one of his major influences. In the official 1853 program for the teaching of Philosophy at the University of Havana, Lamarck appears mentioned again, but this time his work is related to the formation of progressive generations and the process of species formation. We go from "sponge to man, and not viceversa," confirming the hierarchical notion of natural organization (68). But it is by the last decade of the century, when applied to the organization of human societies, that Lamarckian views would become most predominant among Spanish American intellectuals. We can see this turn in the issues of *Revista Cubana* published in 1889. In them there are several articles, including one by Herbert Spencer, in which the discussion focuses around the importance of soft-inheritance in Darwin's work, or the possibility of passing acquired traits to the next generation. In one lecture to the Cuban Anthropological Society in October of 1888, Aristides Mestre explains the connection between anthropology and "scientific politics [política científica]," the example for this relationship being the analysis of colonial society (511). According to Mestre, Darwin was very useful in understanding the evolution of modern colonization, since through his work we come to understand the process of "acclimation," which is "nothing other than a special case of adaptation" (511). Darwin is not here identified with natural selection or the survival of the fittest, but with an adaptationism much closer to neo-Lamarckian ideas. Mestre insists, using Darwin's support of soft-inheritance, that the Iberian people met "special conditions for adapting to the tropics, conditions that distinguish the Spaniards" (512).

This association of Darwin with Lamarckian ideas, also represented by Herbert Spencer, was certainly a recent development. Mestre recognizes that even when not all aspects of the Darwinian theory were fully understood, now "a synthetic and generalizing tendency" prevailed in both the biological and physical sciences. "From all current theories an inclination to unity is deduced," reverting the previous decades' divisions in terms of understanding the goal of evolution (513). Mestre also alludes to other debates about the consequences of racial mixing, and raises questions regarding the influence of the environment. In 1892

Eugenio Sánchez de Fuentes y Peláez, Professor of History at Havana University, wrote in his book about Spanish history that the differences between Lamarck and Darwin consisted in the fact that for the French naturalist the "starting point of transformation resides in the exterior, which in the long run modifies the way of living, creating new needs and customs;" while Darwin's starting point "could simply be found in the superiority that always obtains in the struggle in which one possesses any advantage" (67).

In the United States, another region from which Martí drew many of his sources, this division was also public and very well known. In *The Standard Natural History*, published in 1884, the editors acknowledge their belief "in a modified and greatly extended Lamarckianism, or what may be called neo-Lamarckianism" (liv). While Darwin was right on some issues, his theory could not explain all the instances of evolution, so this new approach had begun to focus on the environmental factors in evolutionary process. W. M. Wheeler explains this shift in a lecture of 1898:

> ...so Darwin's followers have split into two schools. The neo-Lamarckians and the Neo-Darwinians, in obedience to the two psychological tendencies to which I have called your attention. The neo-Darwinians, in laying great stress on the segregation, stable and complex *intrinsic* structure of the germ plasma and its importance as a vehicle of heriditary characters, and in atributing less value to the *extrinsic* factors, like food and the environment, are allying themselves with theorists of the type of Parmenides and Plato. On the other hand, the neo-Lamarckians who believe in the permanent change-producing effects of the *extrinsic* factors (environment, etc.) on structure, and attribute less value to the architecture of an *intrinsic* vehicle of heredity (germ plasma), range themselves with Heraclitus and Aristotle. (282)

During the last decade of the nineteenth century the multiplicity of evolutionary views began to coalesce into two camps, divided over the possible transmission of acquired traits and the influence of the environment on the process of speciation. So, for Latin American intellectuals this meant that there now was much more clarity on how to approach this subject. Minimizing the role of natural selection was the perfect solution for cultural evolutionists, such as Martí. Ideas and culture adapted to the environment to create different evolutionary outcomes. I think that in this context we can understand "Our America" as a neo-Lamarckian text, one that proposes that in America the affirmation of European ideals could only happen after they were naturalized by the influence of the environment. In fact the optimism exhibited in "Our America" can be related to the prospect that the alledged sickness afflicting the post-colonial world had finally found a remedy.

If before the frustration Martí felt about Darwinism derived from the fact that relying on natural selection disempowered human will, and thus any prospect for controlling the pace of social or political change, neo-Lamarckian ideas offered him a different perpective. The incorporation of foreign ideas depended on the ability of the latter to adapt to a new environment. They had to change, and in the process so did the population. The "new American man" was now the result of a process of adaptation and not of replication. In this evolutionary narration everything seems to be in a state of in-betweeness, in a process of negotiation that reunites aspirations and natural conditions. This is process of forming the new generations; but in this evolution there is no racial hatred "because there are no races." The soul emerges, equally and eternally, from "bodies diverse in form and color" (OC VI, 22). "Our *mestizo* America" is a reality, and the mestizo is the symbol of the process of social evolution itself. *Mestizaje* is in this way an indication of posible existences, not a racially determined category. Leaving behind the territory that had fixed racial types, the figure of the *mestizo* represented the beginning of a new stage. But in the importance of the environment, in the emphasis that the text gives to the process of adaptation, there is a clear connection with the French transformist conception of *milieu*, discussed above in relation with Taine. There is here a mechanicism related to the workings of the environment together with a vitalism that comes from the aspirations of the individual, a characteristic related to French Lamarckian thinking.

Following Georges Canguilhem, Paul Rabinow explains that the transformist French notion of milieu had important scientific consequences. First, this concept "resulted in a fundamental decentering of the understanding of nature and society. There was no longer a fixed center –given by God and Nature- capable of being represented." Furthermore, the milieu was no longer "a preordained place, but simply the "between" of two places, *mi-lieu*, a relational system without metaphysical grounding" (128). This process led to "a persistent dualism in French biology. Lamarck proposed a mechanistic conception of the milieu and a vitalist one of the organism. One was basically passive and the other active; although they continually interacted, a fundamental gap remained between them" (128).

In Martí's treatment of the idea of environment we can see similar elements to those described by Rabinow. In addition, he naturalizes the subjectivity of Indians and those of African ancestry, who, according to him, are solely motivated by their desire to change, to improve, to be better. In his view they are waiting to be rescued, which explains why he talks about "mute Indian masses" (OC VI,16). They symbolized passivity (nature) that needs to be activated in order to bring them up to contemporary speed. Silence can only be broken in the expression of displeasure at the fact that their desires to be well-governed, to be transformed, are not taken into account. Of course, the historical inaccuracy of this assertion is quite remarkable. Martí ought to have known that many revolts by the Indians, for example, were not about moving forward, but about defending their right to return to the past. In any case, their supposed passivity is contrasted by the active forces of the elites who want to integrate the nation in a definitive way (the organism). *Mestizaje* reflects then this complex negotiation, in which all participants are willing to transform themselves in the name of the nation.

Julio Ramos invokes Deleuze and Guattari to analyze the symbolic use of the image of a tree in the text, but I think that in using this image Martí is simply refering to an evolutionary commonplace (257). The tree of life had become, particularly after Haeckel, a constant trope of those who followed Darwinian ideas. Martí uses the analogy of the tree and the branches to explain the new articulation of "our America." Employing the language of a naturalist he affirms that what was needed was to "let the world be grafted onto our republics; but the trunk (tronco) had to be our own" (18). The influence of neo-Lamarckianism allowed Martí to contemplate the possibility of cultural adaptation, so the republics that had been declared dead and extinct acquired here a revitalization through the grafting of new life. Earlier I discussed how Martí construes the Spanish American problems in terms of finding "an analogy" between the actual and anticipated worlds, and I think the image of the tree shows that he resolved this problem to his satisfaction. Nature and desire are integrated in this type of analogical thinking.

There is at the same time unity and variation, but the latter is directed by the forces in which the external element is placed. There are differing opinions, but they exist "in one single chest and with one single mind" (21). Everything is a dialogue between inside and outside, with an emphasis on a process of grafting that indicates that whatever is perceived as different can be incorporated without losing the "natural" characteristics. In warning about the perils of incorporating cultural forms without placing them in dialogue with the internal, real life of a society, he is expressing neo-Lamarckian concerns about adaptability. According to Ian Robert Dowbiggin, in French neo-Lamarckianism the "characteristics that led to successful adaptation depended on the ability of the organism to adjust to changing environmental conditions" (148). In this context, not all European ideas were able to adapt to the American environment, and in order to discover which ones could, they had to enter into a process of negotiation with the "natural man." The result would be, as in a grafting, that the characteristics introduced would be processed by nature, which will adapt the foreign element according to its own needs. Grafting, after all, is a process in which difference becomes visible, but without tension, since it only reveals a new possibility, a new variety, of the being onto which the graft is inserted. More than anything this process is one that is predicated on the possibilities of existence, on being natural at the same time that natural patterns are being broken.

In the naturalist's analogies that are crucial in "Our America" we can see how Martí retains a sense of a duality, but this time systematized around evolutionary ideas. In this way, the tree (the species) is the real, while the variations are the ideal. As if intellectuals were gardeners, Martí assumes that not all the ideals grafted onto a species can work; one must await nature's reaction to observe the results, but in order to ancitipate this reaction we need to fully understand the environment in which this particular species has evolved. But if the adaptation is a success, and the ideals are realized, then what was changed is rooted in nature, and is part of a new origin that is at the same time authoctonous and foreign. It is in this resolution that he incorporates an idealist vision into his evolutionary tale. Authoctonous refers here to the origins, to the natural environment in which a given species, in this case the 'natural man,' emerged. But the process of mixing reflects man's aspirations as negotiated (made real) in the context of stable natural mechanisms. In this sense, Martí reproduces a basic Darwinian logic. It is important to remember that for Darwin binary divisions do not indicate that one overpowers the other, but that there is a coexistence of life/death, evolution/extinction, and continuity and discontinuity.

Conclusion

> No sé si soy loco, puesto que soy un idealista tan completo. El realismo santo maravilloso, milagroso, es la lógica de la naturaleza. [I do not know if I am crazy--I'm such a thorough

> idealist. Blessed, marvelous, miraculous realism is the logic of nature.] Martí, 1879. (OC XIX, 429).

"Our America" is one of the most important texts for the discussion of identity in post-colonial Spanish America. In this essay, I have tried to show how among all the different meanings that such an important work can have, there is also a scientific connection that is worth exploring. Martí's interest in science and his critical analysis of Darwinian ideas remind his readers to avoid promoting a theory that was politically so close to the foundations of imperialism, and the division of humankind into winners/losers, selected/discarded. In his desire to promote a vision of a nation that was inclusive and based on an inter-subjective positionality that favors racial integration, he refuses to accept the work of Spencer, or the idea that the Anglo-Saxon was the race of the future. Even when Martí's thinking was developed in the context of a culture dominated by evolutionary debates, he did not react like his Argentine counterparts Octavio O. Bunge and José Ingenieros, for example. While the latter embraced the ideas of racial progress based on the Darwinist currents of the time, Martí called the attention to the relationship that scientific views had with cultural and political interests. I agree with Ofelia Schutte when she characterizes Martí's vision as integrative, a tendency that is evident in most of the works that influenced him (147). This strong interest in preserving a sensibility that promoted integration is what explains his adoption of an adaptationist evolutionary logic in "Our America" .

In his essay on social psychology, written in 1903 and also entitled *Nuestra América*, Carlos O. Bunge wrote that in order to understand the uncertainties of life we need to study "men and peoples according to race," which made him conclude that *mestizos* were an uncertain type whose development was doubtful (99). In the case of Martí, his evolutionary thinking does not associate race and nation as a single unit that has to develop together. For him, then, a "*mestizo* America" is not a racial category, but an evolved way to understand the world, which in the scientific language of the time meant a psychology developed through the struggle between ideas and environment over a given span of time. It is no wonder that today Martí's work is widely known, while very few recognize the version of *Nuestra América* discussed by his Argentine counterpart. I believe that studying Martí's development in connection with evolutionary ideas makes very clear his dilemmas between accepting a science that he thought correct at the cost of sacrificing his political ideas. His final adoption of some evolutionary principles through adaptationism came after the science itself had changed and developed a synthetic unity abandoning its emphasis on racial competition.

On the other hand, when analyzing his cultural evolutionism, as it is expressed in "Our America," there is another side that emerges. In it the interest in unity seems to get in the way of a more inclusive society. In its systematization of evolutionary ideas, Neo-Lamarckianism allowed a scientific view that restored important beliefs inherited from pre-Darwinian days. The nation could be unified again, progress was a benign force that could benefit all individuals as long as they aspired to improve themselves and were able to adapt to changing conditions, and a faith in progress was again at the center of evolution. Martí's embracing of adaptationism, I believe, is related to this important transition. As the analogy of the tree that needs grafting seems to indicate, the "natural man" needs to be diversified, a naturalization that refers to the incomplete 'nature' of the species. The grafting only indicates that the cultural diversity of Spanish America is a problem that needs to be resolved. The analogy he takes from horticulture—a kind of artificial selection–and the adaptation of species only serves to underscore the fact that the realization of the ideal comes from the filter of nature, which makes it real. So, in this process he opens the possibility to enjoy the multiplicity of colors (variety) under one single way of perceiving them (unity). Even when this sense of oneness comes from a dialogue between the inner and outer elements that make the nation, it is clear that the final result will not be multiple, but single. As Martí clarifies towards the end of his essay, the duty of showing the post-colonial reality as it is reveals that it is "one, in soul and intention [alma e intento]," (22) and a "unanymous hymn" announces a new time (23). This time unity is not embodied in a single race, but in a single way to comprehend reality. But agency being one of the crucial elements re-instated by social-Lamarckianism, we can ask whose agency this is, how Martí's naturalization of a subjectivity that promotes progress and evolution can provide the results that he honestly expects.

He clearly critizices the Eurocentric fever that led some leaders to think that any idea, any concept coming from Europe could be adapted to Spanish America regardless of the conditions of existence. On the other hand, it is clear that he shares some of this eurocentric attitude when "Our America" is so devoid of concrete examples that make visible the subjectivity of that other whom he sincerely wants to integrate. When he calls for the teaching of the history of ancient American civilizations, he is doing so with a view toward discovering "who we are," which also means addressing the problem of the pasts living inside us. Is Martí suggesting that we need to learn about these civilizations so we can be like them? It seems to me that this is not the case. His text does not suggest a view in which Aztec life could be revived. But in remembering it, in recognizing it as a thing of the past, a previous manifestation of a form of being, he is making sure that variation will not be lost, that it will be preserved as, the most essential element of a Darwinian conception of culture, in which everything is at the end a museum piece.

WORKS CITED

Programa Oficial de las Materias Concernientes a las Distintas Asignaturas de da Facultad de Filosofía de la Real Universidad De La Habana. La Habana: Imprenta de M. Soler, 1853.

Aitken, Stuart C. *Geographies of Young People*. New York: Routledge, 2001.

Arango, José Francisco. "Consideraciones Sobre el Espiritismo." *Revista de Cuba* VIII (1880): 32-43.

Bunge, Carlos O. *Nuestra América. (Ensayo De Psicología Social).* 1903. 2nd ed. Buenos Aires: Valerio Abeledo, 1905.

Camacho, Jorge. "Signo De Propiedad: Etnografía, Raza y Reconocimiento en José Martí." *A Contra Corriente* 5.1 (2007): 64-85.

---. "Los Caminos Trillados". *Opinión*. Cubaencuentro. 1/12/09 2008. <http://www.cubaencuentro.com/es/opinion/articulos/los-caminos-trillados-89952>.

Casas, Juan Bautista. *La Guerra Separatista de Cuba. Sus Causas*. Madrid: Est. de San Francisco de Sales, 1896.

Catalá, Rafael. "Ciencia y Literatura: Su Función en los Pueblos de América." *EducAcción* 6 (1986): 4-6.

Di Gregorio, Mario *From Here to Eternity*. Göttingen: Vanderhoeck & Ruprecht, 2005.

Dowbiggin, Ian Robert. *Inheriting Madness*. Berkeley: University of California Press, 1991.

Frías y Jacott, Francisco de. "Sobre *El Origen de las Especies*." *Revista de Cuba* VIII (1880): 213-20.

---. "Cartas." *Revista Cubana* 2 (1885): 538-41.

Gellner, Ernest. *Culture, Identity, and Politics*. Cambridge: Cambridge University Press, 1987.

González, Aníbal. *Journalism and the Development of Spanish American Narrative*. Cambridge: Cambridge University Press, 1993.

---. "Literary Criticism in Spanish America." *The Cambridge History of Latin American Literature*. Eds. Roberto González Echevarría and Enrique Pupo-Walker. Vol. 2. The Twentieth Century. Cambridge: Cambridge University Press, 1996. 425-58.

Guerra, Lillian. *The Myth of José Martí*. Chapel Hill: University of North Carolina Press, 2005.

Hale, Charles. "Political Ideas and Ideologies in Latin America, 1870-1930." *Ideas and Ideologies in Twentieth Century Latin America*. Ed. Leslie Bethell. Cambridge: Cambridge University Press, 1996. 133-207.

Hunt, Michael H. *Ideology and U. S. Foreign Policy*. New Haven: Yale University Press, 1987.

Huxley, Thomas H. The Origin of Species." *Revista de Cuba* VIII (1880).

Hyslop, James H. "Why Are We Imperialistic?" *International Journal of Ethics* 16.1 (1905): 1-15.

Jimenez, Luis A. "José Martí y el Comportamiento de los Animales en *La Edad De Oro*." *Ometeca* 3.2 (1996): 250-59.

Jordan, David Starr. *The Days of a Man. Being Memories of a Naturalist, Teacher and Minor Prophet of Democracy*. Vol. 2. New York: World Book Company, 1922.

Jorrín, Miguel. "Martí y la Antropología." *Antología Crítica de José Martí*. Ed. M. P. González. Mexico: Editorial Cultura, 1960.

Kingsley, John Sterling, ed. *The Standard Natural History*. Vol. I. Boston: S. E.

Cassino and Company, 1885.

Manzo, Kathryn A. *Creating Boundaries. The Politics of Race and Nation.* Boulder: Lynne Rienner, 1998.

Martí, José. *José Martí. Obras Completas.* 26 vols. La Habana: Editorial de Ciencias Sociales, 1991.

Martínez Villergas, Juan. *Poesías Escogidas de Juan Martinez Villergas.* Ed. Casino Español de la Habana. La Habana: Imprenta Militar de Soler Alvárez y Cia., 1885.

Merchan, Rafael M. *Estudios Críticos.* Bogotá: Imprenta de la Luz, 1886.

Mestre, Antonio. "Origen Natural Del Hombre." *Revista de Cuba* V (1879): 419-528.

Mestre, Aristides. "Memoria Anual." *Revista Cubana* IX (1889): 510-20.

Pedroso, Carlos de. "La Física Materialista De Nuestros Días." *Revista Cubana* 2 (1885): 42-130.

Pruna, Pedro M. "La Recepción De Las Ideas De Darwin En Cuba Durante El Siglo XIX." *Quipu* 1 (1984): 369-89.

---. "Biological Evolutionism in Cuba at the End of the Nineteenth Century." *The Reception of Darwinism in the Iberian World.* Eds. Thomas F. Glick, MIguel Puig-Samper and Rosaura Ruiz. Vol. 221. Boston Studies in the Philosophy of Science. Berlin: Springer, 2001.

Rabinow, Paul. *French Modern.* Boston: MIT Press, 1989.

Ramos, Julio. *Divergent Modernities.* Trans. John D. Bianco. Durham: Duke University Press, 2001.

Renan, Ernest. "What Is a Nation?" *Nation and Narration.* Ed. Homi K. Bhabha. New York: Routledge, 1990. 8-22.

Rotker, Susana. *The American Chronicles of José Martí.* Trans. Katherine Semler and Jennifer French. Hanover, NH: UPNE, 2000.

Sagra, Ramón de la. *Principios Fundamentales para Servir de Introducción a la Escuela Botanica-Agricola del Jardín Botánico De La Habana.* La Habana: Imprenta de La Amistad, 1824.

Saldívar, José David. *The Dialectics of Our America. Genealogy, Cultural Critique, and Literary History.* Durham: Duke University Press, 1991.

Sánchez de Fuentes y Peláez, Eugenio. *Historia Crítica De España.* La Habana: La Poesía, 1892.

Sapp, Jan. *Evolution by Association.* New York: Oxford University Press, 1994.

Schutte, Ofelia. "Nuestra América en las Postmodernidades Americanas." *Pensares y Quehaceres* 2 (2006): 145-55.

Shideler, Ross. "Questioning the Father: From Darwin to Zola, Ibsen, Strindberg, and Hardy." Palo Alto: Stanford University Press, 1999.

Spektorowski, Alberto. "The New Right and the Intellectual Elaboration of Ethnic Europe: From Racism to the Politics of Differentialism." *Ethnic Challenges to the Modern Nation State.* Eds. Shlomo Ben-Ami and Yoav Peled. New York: Macmillan.

Stoetzer, O. Carlos. "Krausean Philosophy as a Major Political and Social Force in the Modern Argentina and Guatemala." *Bridging the Atlantic*. Ed. Marina Pérez de Mendiola. New York: SUNY Press, 1996. 83-107.

Torre, Carlos de la. "La Estatua de Lamarck." *Revista de la Facultad de Letras y Ciencias* 5 (1907): 280-88.

Wheller, W. M. . ""Caspar Friederich Wolff and the Theoria Generationis." *Biological Lectures from the Marine Biological Laboratory Wood's Holl, Mass.* Boston: Ginn & Company, 1899.

CHAPTER 10

Adaptive Functions of Origin Tales: Chican@ Culture and the Aztlán Myth

Story-telling is a human universal, common to all known human cultures (Brown 132). Many human groups include an origin myth or tale in their narrative repertoire. As with other cultural commonalities, such universality and ubiquity indicate that origin myths probably serve some adaptive function in an evolutionary sense—that groups and/or individuals who recycle a story of their origins survive and reproduce better than those who don't. But why would any story, let alone one about a particular tribe's place or process of origination, have such adaptive valence—particularly when virtually every origin myth is riddled with fabrications?

In the 1960s the Chican@ (the term is used to include both the masculine and feminie forms of the word in Spanish) culture of the Southwestern United States resurrected and began to deploy an origin myth of the ancient Aztecs, the Central Mexican culture that Hernán Cortés encountered and eventually conquered, when in the early 1500s he invaded the lands that would become Mexico. Modern Chican@s have used the Aztlán myth in an attempt to unify the disparate and dispersed people who are the offspring of European and Native American intermingling in Mexico and the American Southwest in the half-century since Cortés's invasion. While this appropriation has come under considerable criticism since it was first resurrected in the 1960s (see Bebout, Alarcón), the Aztlán myth continues to be invoked in scholarly papers and at professional conferences, though for various reasons. This utilization is a complex and ambiguous extension of the traditional function of origin myths. Whether it ultimately helps or hurts the Chican@ movement, it is an interesting and provocative experiment in adaptive story-telling.

The human relationship with "home" is also ambiguous. Human beings, like other territorial animals benefit from establishing and maintaining a home "turf" (Brown; see also Casimiri, Wilson). Yet mankind is also a peripatetic species. All current information indicates that we arose as in East Africa about 150,000 years ago, and left that, our common homeland, about 50,000 years ago, filling every corner of the globe, and eradicating whatever other hominid species we encountered in the process. Implicit in this rapid expansion was the fact that few species subgroups were resident in any particular place for long. Archeology and DNA studies show that this process was actually a series of expansions; that the earth is a palimpsest of multiple human migrations, one group moving into another's territory and either expelling, displacing, absorbing, integrating with, or destroying the other group in the process, over and over again, for the last fifty millennia. Few if any of the current resident populations around the planet occupy lands that weren't previously occupied by other groups (certain east Africans and the Australian Aborigines are rare but notable exceptions). Yet many "indigenous" human groups have origin myths which indicate that they are the rightful and/or first occupants of their homeland. Clearly such claims are not true. But why are righteous origin myths so common? One way to answer the question of why such myths are so often generated and repeated is to look into human instincts.

Part of the reason we tell origin myths is that belonging to *somewhere* is important to us. Anthropologist Paul Basu cites an informant, a Scottish emigrant to Australia, discussing the Scottish diaspora and the case of the Australian Aborigines after the British occupations of both countries:

> I have noticed the similarities between what happened to the Highlanders in the [Highland] Clearances and what happened to the Australian Aborigines (i.e. a whole load of men in red coats arrive and say, "get off the land, we want to put sheep here!"). Unfortunately, I'm not sure this made my pioneer ancestors any more sympathetic to the

> local Aborigines. It seems to have been a case of "what's done to you, you do to others". But I do think for us in the diaspora ... our idea of "homeland" is inextricably linked to the place we live in now. And I sometimes wonder if part of the longing for a homeland is a longing for a place of ... belonging, especially in the current Australian climate ... The idea of a place where my ancestors belonged for thousands of years is very appealing (338).

The need to belong may derive from our territorial instincts. Territorial behavior in humans appears to be an instinctive, inheritable capacity that is generally context-dependant—it arises in certain environments, lies dormant in others (cf. Buller 46-7; Casimir). The evidence that substantiates claims of human territorial behavior derives from a number of environmental facts. Among territorial animals, those who establish and defend a territory have better chances of surviving and reproducing than animals that don't. Food resources are more under their control. The lay of the land is better known, so defensive and aggressive logistics are better understood. Predators and rivals are more easily driven away. Offspring are protected. In the larger scheme of things, social structures are maintained, and an apparently associated instinctive behavior—acknowledgment and respect of territorial rights by outsiders—actually reduces social conflict within breeding groups, controls population, and conserves valuable resources: in this sense, territorial behavior is good for both the proprietor and its social group. Territorial behavior seems at root to be a response to food resources, and where such resources are either overabundant or mobile (fish or winged insects) territorial behavior is often non-existent.

Human beings happen to be primates, a group of animals among whom territorial behavior is common. In their provocatively entitled book *Demonic Males, Apes and the Origins of Human Violence*, primatologists Richard Wrangham and Dale Peterson, both benefiting from extended field work alongside Jane Goodall, describe territorial behavior and its occasionally disastrous consequences among groups of African chimpanzees. An assumption behind their book, one substantiated by their research, is that territorial behavior is part of our inheritable repertoire, but that such behavior is quite malleable, and usually environmentally driven.

Anthropologists continue to discuss territorial behavior in a variety of manifestations. For example, Wrangham and Peterson include physical force as part of their definition: "Territory is a range that is forcibly occupied" (14n). These two writers mollify sensibilities by adding that a "range ... is a piece of land that is occupied whether or not force is employed." (Ibid.). The fact that their book is about violence may explain the distinctions in their definition—they are writing to offer solutions to problems stemming from human violence, and physical force thus plays a role in most of their discussions. But for both "territory" and "range," more-or-less exclusive occupation is the common characteristic in their definition. On the other hand, sociobiologist E. O. Wilson's definition is less rigid: "Territory [is] an area occupied more or less exclusively either directly by overt defense or indirectly by advertisement. This area invariably contains a scarce resource ..." (110). This definition is more concerned with ends than means, and exclusive occupation is still the immediate goal.

Foundational ethologist Irenäus Eibl-Eibesfeldt's definition is yet broader, hinting at the plasticity of the territorial "disposition":

> Man is also disposed to take possession of land and to delineate between himself or his group and other individuals and groups. Group members respect the territorial claim of another group member as his possession, as if following some primary norm involving respect of possession. These moral inhibitions against territorial trespassing are less defined if the potential intruder does not know the territory owner. Fear of retaliation then becomes the stronger force to keep the potential intruder from trespass and conquest . . . As long as threat of intrusion is absent, one often does not notice any manifestations of territorial ties to an area. But once such a threat appears, the social relationships of ownership in the group are projected onto the land . . . In

> most cases, human group territories are protected by the mere knowledge of the possessors' readiness to defend the territory, with force, if necessary (321).

In turn, French anthropologist Michael J. Casimir's definition further emphasizes such flexibility:

> Human territorial behavior is a cognitive and behaviorally flexible system which aims at optimizing the individual's and hence often also the group's access to temporarily or permanently localized resources, which satisfy either basic and universal or culture-specific needs and wants, or both, while simultaneously minimizing the probability of conflicts over them (20).

But all writers on the subject seem to agree that territoriality is an innately human behavior. Of course, the question of whether human beings tend to be territorial can be answered by anyone with a daily paper or a television, who grew up in a neighborhood, went to a school, worked in an office, etc. Casimir's definition above is drawn from a text which discusses the territorial behavior of nomads and peripatetic tribes—people traditionally defined as having *no* exclusive territory, yet who manifest territorial behavior in certain circumstances (Casimir and Rao). But the question of whether the tendency is inherited is a bit more complex. Anthropologist Julian Steward offered evidence towards an answer in 1936, when he published early studies showing that two dozen remote, disparate human groups with no chance of inter-communication all possessed and defended exclusive territories. In 1991, Donald Brown confirmed this in his exhaustive book, *Human Universals*, with territoriality as number 147 in his list of 155 traits common to all human cultures ever studied (Brown). Given this ubiquity, the most likely explanation is that territoriality constitutes an inheritable predisposition that is a functional adaptation to most environments found on our planet.

If for adaptive reason some affinity for native territory is inherited in most human beings—if in all humans and many animals it is a universally preferred tactic for exploiting terrestrial environments, for surviving and reproducing—it makes sense that the loss of a territory provokes stress. Nature works by eliciting emotions of pleasure and displeasure, with pleasure acting as the carrot, tempting us towards what has traditionally been good for us, while a variety of displeasures are the stick, driving us away from what has traditionally been bad. Sugar gives us pleasure because it is high in calories, and high-calorie foods have helped mankind survive across time. In psychologist John Bowlby's work on separation anxiety, neonates experienced stress—displeasure—from prolonged absence of their primary care-givers (Bowlby). In humans, as in many other species, the absence of a primary care-giver equals death. Hence the distress. Foods that give us pleasure usually do so because they increase (or increased in the past) the ability to survive. Loss of native territory causes displeasure because it reduces our ability to survive.

Various field studies offer empirical evidence of this stress. British anthropologist Colin Turnbull's study of the Ik, the so-called "Mountain People" of Uganda, claimed that when the tribe was forcibly moved to make way for a park, their social structure collapsed and interpersonal morality completely degenerated. Turnbull's study and methods have been heavily critiqued in the interim, but the basic facts—forcible removal from territory, followed by societal collapse—remain. Turnbull's study finds incidental corroboration in a later, more rigorous study, of the Ju/Wasi of Nyae Nyae, a tribe that lives along the border of Namibia and Botswana in southern Africa. The Ju/Wasi (translation: "the well-mannered people who speak correctly") may be one of the last hunter-gatherer peoples on the planet. In the last half-century, government machinations on different fronts have altered their territory and lifestyle. By mere happenstance, the Ju/Wasi were separated into two groups living on neighboring territories—one residing on the original home territory, with known water and other food resources, while the other resided on a adjoining desert where no Ju/Wasi or other human group had ever lived. The home dwellers were allowed to retain some of their original foraging and hunting behavior, though over time, a conscientious government officer helped them to add subsistence farming and pasturing, along with some selling of goods, to their lifestyle. The transplanted desert dwellers could not follow their traditional way of life, because they had been moved from their home territory onto properties with no water or viable food sources. This second group lived off government handouts and employment with the Namibian army. From 1970 to about 1984 (the time of the publication of the study), the health and well-being of the two groups diverged drastically, with the western home dwellers developing a multi-faceted economy and expanding their population, and the eastern desert dwellers degenerating into a welfare economy, with high rates of alcohol abuse and crime. Ironically, both groups are now threatened by the forced relocation of other, non-Ju/Wasi Bushman peoples into already crowded portions of their lands, and by government attempts to turn the "home dweller" portion into a nature preserve, where many traditional subsistence methods will be banned.

In a letter to the Namibian government about these proposals, a Ju/Wa elder speaks volumes:

> We are Ju/Wasi
> We do not want a nature reserve …
> People are growing old, and they do not know where they will go … and the government wants the Ju/Wasi to leave their land with their cattle and go to a strange land ...
> And people's hearts are sore because people will be destroyed
> We will stay here
> Because this land is Ju/Wasi land. And it is a little bit good …
> And we know this land, and do not know strange places.
> We know where water is. And bushfoods …
> We will stay (here) where God himself made the water.
> I am finished (Marshall and Ritchie 12-13).

The Ju/Wasi know that moving away from their tribal lands is a bad idea. This leader's iteration of the reasons is eloquent in its simplicity.

Yet claims that intransigent territorial behavior is neither adaptive nor innate offer plenty of empirical justification. Human beings wander, frequently migrating away from their birth places, frequently marrying outside their tribe, and often surviving and reproducing spectacularly in all of these efforts. Territorial behavior is a cultivation of the norm, where emigration is an attempt to exploit the novel. Both have adaptive valence, but the quotient of risk in each case is quite different. Where the norm is simply the norm—familiar faces, routine sounds, predictable resources, etc.—novel occurrences can represent potential threats or potential opportunities. An unrecognized face on the periphery of the group territory can represent a threat, a mating adventure, a warning, a trade opportunity, etc. Avoiding unusual threats has obvious survival benefit. Exploitation of novel resources can provide a competitive advantage. And unexpected mating opportunities are the proverbial brass ring in proximal adaptation strategies.

Psychological studies of individual human "life histories" have shown that people assess their birth and natal context quite early in life, making decisions about whether cultivating the norm or attempting to exploit the novel will have the best potential for future pay-off. In Scandinavian cultures primogeniture—the inheritance of leadership and property prerogatives by the eldest son of a family—was practiced historically and eventually legally substantiated, so that sons subsequent to the first often found themselves in a situation where adherence to natal territory limited their future. Consequently they struck out for new horizons. This is a good example of the "life history" lesson, that leaving might be risky, though staying might be worse. In turn, women have routinely left their birth lands and tribes for various reasons. From a genetic standpoint exogamy brings clear benefits, with incest avoidance reducing repetitive genetic mutations in children. Multiple studies have demonstrated that it is instinctive in human beings to avoid mating with close relatives. Inter-group marriage has also played a role in reciprocal social relationships (cf. ***Beowulf***, not to mention, the history of the European monarchies). The subsequent treatment of fertile women as chattel and commodity has been extensively discussed by writers in various fields. Whole cultures can also benefit from emigration. Imperialism is often a case study for exploitation by emigration, although this pattern is only the latest example of the sort of habitual, often exploitive, emigration that has characterized human behavior across time. Most of these emigrants set up territories when they make landfall. And they often then compose a story justifying their "emigration."

In Wilson's definition, above, it is clear that monopolization of necessary resources is the initial and driving force behind territorial behavior. His brief list of possible resources includes "steady food supply, shelter, space for sexual display, or a site for laying eggs" (110). He adds that territorial defenders are practically invincible within their territories, regardless of their defensive or aggressive prowess outside that territory (ibid). This suggests some of the value of a "home turf." Here Wilson was referring in generic terms to general territorial behavior, but other authors develop the concept of territory in humans, to the point where personal and group identity is intimately intertwined with the land (cf. Lovell, *Territory and Belonging*). This adds identity to the group of things being isolated and defended—the things potentially lost when territory is lost.

Clearly this theoretical framework has some relevance to the Chican@s—a human subgroup that has been radically deterritorialized. But territorial disenfranchisement has other side effects that may be equally relevant.

Another consideration accompanies this correspondence among natives, their native identity, and their native land. Our need to belong may also derive from another set of instincts, instincts that incline us to belong to some *group*, rather than just to some piece of property. Those who live in "our" territory or group are "Us," and those who don't are "Them." Distinctions between ingroup and outgroup are extremely important to human survival. A person who leaves his or her native land almost automatically becomes an "outgroup" member, an automatic "Other," as the result of the migration or expulsion. And that exclusion is usually not a good thing.

An extensive quotation from Wrangham and Peterson makes this point:

> People quickly form groups, favor those in their own group, and are ready to be aggressive to outsiders. We all abhor such biases. They lead to all sorts of "isms"—racism, sexism, ethnocentrism, and so on. But we all fall prey to them with appalling ease …
> The temperamental complex involved is what we call *ingroup-outgroup bias*. The ingroup-outgroup bias is often ethnocentric, which means that the ingroup and outgroup are perceived as different races or ethnicities, but it can appear with equal ease around other categories, such as religion, sex, age, or sports team. In striking contrast to many or most processes they describe, social psychologists describe this process as *universal and ineradicable* [emphasis added].Taken to an extreme, ingroup-outgroup bias effectively dehumanizes Them, which means that moral law does not apply to Them and that therefore even ordinary and very moral people can do the most appalling things with a clear conscience (195-96).

The authors go on to cite horrific examples from Manifest Destiny, the slave trade and, interestingly for our particular topic, the Spanish conquest of Mesoamerica.

To that same point, Ian Jobling, a professor of comparative literature, has done an extensive study of hero/ogre stories from forty-four separate cultures, and concluding: first, that this story line is far too common to have been culturally transmitted; second, that the tendency to construct and consume such stories must derive from something innate in human nature; and, third, that the function of such stories is to "reinforce innate positive biases in the perception of self and ingroup and negative biases in the perception of outgroups" (1). Jobling believes that we have an innate—not culturally transmitted, but resident in our genes, formed there during the process of our evolution—predisposition to tell tales in which a heroic member of "Us" defeats a "semi-human ogre" who is a member of "Them." The need to cultivate ingroup/outgroup distinctions figures as a deep legacy in our species. The ubiquity and the universality of the plot structure suggest that the distinction between Us and Them served adaptive purposes, with a survival value that over time has been selected for by the evolutionary process. The fact that the Us member is invariably written as a hero, and that the Them member is inevitably written as a subhuman to be destroyed by the hero, along with abundant human histories where noble "Us" tortures, rapes, kills, and dispossesses subhuman "Them," makes it pretty clear why the distinction between Us and Them has been important historically, with implications for our prehistory. And pity Them.

Again, these constructions are intuitively sensible and empirically obvious. Outgroup members, the Others, are invariably more vulnerable than ingroup members. Evolution should thus favor organisms with a predisposition to retain ingroup status and avoid outgroup dangers. Yet most territorial dislocations mandate a change to outgroup status. When you leave your native territory, for whatever reason, you are almost automatically an outsider in the new territory. Even without the territorial argument advanced above, this change can be an obvious source of stress and potential trauma to the dislocated.

Embedded in the material from Wrangham and Peterson concerning ingroup/outgroup biases is the claim "We all abhor such biases." But do we? Jobling cites Levine and Campbell's study showing that "ethnocentrism, or positive bias in the perception of ingroup members and negative bias in the perception of outgroup members, is a universal human phenomenon" (252). Further:

> People universally see the ingroups to which they belong as "virtuous and superior" (Levine and Campbell 1972:

> 13) and "strong" (1972:14). Moreover, people tend to "see [the] outgroup as contemptible, immoral and inferior" (1972:13) and consider the killing of outgroup members in warfare as virtuous (1972:18) (252).

Wrangham and Peterson's statements serve as instantiation of Levine and Campbell's claim—"we" all abhor such negative outgroup biases because we are virtuous: yet we all, according to the evidence of the cited study, harbor such biases. And, occasionally, when we want to emigrate to another group's territory, we label them outgroup/Them/Other, contemptible, immoral, and inferior, and may even consider killing them to be a virtue. And how do we accomplish this hypocrisy? Jobling says that we (all human cultures) narrate a Hero-Ogre story about Them.

Of course, we respond differently to voluntary dislocations, to moves under our discretion and control, than we do to involuntary dislocations, deportations, evacuations, expulsions, etc. Psychologists distinguish between our reactions to voluntary and involuntary stress:

> Clearly, the nature of the stressor is likely to determine the degree and quality of its impact. Correspondingly, different stressors tend to evoke different types of responses. For example, *coping* processes are believed to vary according to whether the situation is perceived as controllable or uncontrollable. When it is controllable, direct action designed specifically to correct the stressful aspects of the situation (*problem-focused coping*) is more appropriate, whereas when it is uncontrollable, people may be able only to work on their affective reactions to what has happened and try to make themselves feel better about it (*emotion-focused coping*) (Parkinson and Colman).

It seems likely that we use origin narratives to maintain attachment to territories, and that journey/origin narratives are an adaptive response, a type of emotion-focused coping, to involuntary loss of territory and the associated outgroup status.

Territorial dislocations are a fact of human history and prehistory, common to the point of being commonplace. Yet the weight of evidence indicates that we have an innate predisposition to acquire and maintain a territory and to avoid forced dislocation; that our ability to survive and reproduce is threatened when we fail to avoid it, and that we experience considerable stress at this failure. There is a conflict here between instinct and empirical reality. We may move a great deal, but involuntary moving is bad for us. For human beings, culture usually arises in to bridge such gaps.

The standard historical model indicates (as do most of the counter-narratives) that some of the Native American ancestors of Latin@ people along the Texas/Mexico border once asserted exclusive possession of property that is now in the possession of people whose ancestors came from Europe. At the time of the Spanish invasion of Mesoamerica, the Náhuatl people occupied Tenochtitlan and neighboring cities in the valley of central Mexico, where Mexico City now lies. Various journey tales and origin myths suggest that the Aztec people came to Tenochtitlan from somewhere else, an ambiguously described territory somewhere north of Mexico City. All of these narratives describe a displaced people, who in most cases are not now the proprietors of the land they live on, and who for the last century or so have been treated as outgroup in the land they occupy. Thus territorial loss and avoidance of outgroup status is one likely explanation for the resurrection of the Aztlán myth.

This specific function of origin myths has been explored in great detail by various social scientists. Anthropologists Barbara Bender and Margot Winer begin their 2001 collection, *Contested Landscapes: Movement, Exile, and Place*, with a dedication to the expropriated, to the sort of forcefully dislocated subjects discussed here: "We dedicate this volume to people-on-the-move, for whom landscapes are never a taken-for-granted, and are often a source of pain" (v).

Real-world examples of the coupling of home, identity, and story-telling are forcing a productive move towards empiricism and away from theory of anthropologists who study the relationship between "home" and personal and cultural identity:

> ... there is a certain irony in the fact that while social scientists have been actively "liberating" the concept of culture and identity from previously assumed bonds to location in place, social groups are actively involved in constructing and defending all of the things under re-examination: rootedness and belonging, ways of life, realities and senses of place, boundaries that certify and maintain distinctiveness (Caftanzaglou 21).

In an essay in the Bender and Winer text, Paul Basu parses the distinctions between narrative identity construction via affiliation with a static homeland, and construction of identity by people who are in motion. Basu describes the two different approaches as "the broadly modern, which allies identity with fixity, and the broadly postmodern, which asserts the ability of the individual to find 'home' in movement" (334). The first, the modern, is the commonly perceived affiliation of person/people with place (cf. Kiberd, *Inventing Ireland;* Lovell, *Locality and Belonging*). The second is more problematic, with one camp emphasizing what Basu calls "the ontological uprooting caused by physical migration and displacement," creating what various theorists call the "homeless mind," which arises when "the anomy of social movement correlates with personal alienation on the level of consciousness":

> ... the "homeless mind" is hard to bear, and there is widespread nostalgia for a condition of being "at home" in society, with oneself, and with the universe: for homes of the past were socially homogeneous, communal, peaceful, safe, and secure (Rapport and Dawson 31).

This theory of homeless anomie is contested by another theoretical camp which finds ethnographic support for the claim that Bender and Winer's "people-on-the-move" are actually "*at home* in personal narratives that move away from any notion of fixity within a common idiom, and their identities derive from telling *moving* stories of themselves and their worldview" (Rapport and Dawson 30; qtd in Basu 334; emphases added). One group sees the displaced as forever wounded, while the other sees the same people salving territorial loss with creative tales of home. These opposing theories link people, homeland, and narrative, in an identity-creating and stabilizing equation, showing that journey and origin narratives work to establish, at least psychologically, residency and rootedness for both the settled and the displaced.

Viewing ethnographies and stories from both "on-the-move" people and emplaced residents on either side of the Scottish diaspora, Basu recognizes the double-edged nature of these stories:

> It seems, therefore, that we ought to address both the mobile and the static in these narratives of identity, and recognize that, while in certain circumstances home may become deterritorialized, in many cases senses of territorial and social rootedness continue to give people ontological security in a world of perceived movement (336).

The idea here is that the concept (not the reality) of "home" provides identity foundation to both "people-on-the-move" and indigenes. Basu's clever metaphor describes these people as "more rhizomic than rooted" (335). His disagreement stems from a perception of the function of journey and origin tales that is very close to my thesis herein:

> I should like to question Rapport and Dawson's assertion [that narratives fail to palliate dislocation]. For, in discussing narrative as a mobile resource, they seem to be discounting the possibility that it may also be used to articulate senses of rootedness. Indeed, they may be correct in doubting the (objective) reality of lost edenic homes evoked in migrant tales ... but surely they ignore the (subjective) reality of these putative homes for the

> people who invoke them. Similarly, they describe contemporary identity as a search (1998:4), but perhaps they do not explore the full range of the metaphor. Is not to search, in part, to seek that which is felt to have been lost? (335)

For Basu, the objective reality—the truth—of origin tales is irrelevant. The displaced story-tellers invent, at least in part, what Rapport and Dawson skeptically call "that 'original life-world' of traditional absoluteness and fixity" (32) in order to fill a particular need, and the "(subjective) reality of these putative homes" serves the same purpose as any objective reality of home. Basu claims that (partially) invented, or partially fictionalized, homelands, are adequate surrogates for actual homes for wanderers—that we react to the invented or mythologized home like we would to a real one. In our minds, the mythical becomes the real.

Basu's discussion of what he terms the "tropes" of "diaspora" and "homeland" is illuminating:

> These are relational terms like "the global" and "the local", each constituted in the other such that there can be no diaspora without homeland (actual or imagined), no homeland without dispersal. Indeed, part of the recent redefinition of "diaspora" (Safran 1991: Cohen 1997) has been not only to recognize the diversity of reasons for population displacement (giving rise to "labor", "trade", and "imperial" diasporas in addition to the archetypal "victim" diaspora) but also to realize how essential the "myth" of homeland is in maintaining senses of common identity amongst such populations (336).

Here the mythologized home becomes a symbol that serves its function in the absence of a real home—the tale of homeland becomes a security blanket of sorts, unifying dispersed populations, as if they were still one people in one place. As is discussed in detail below, this is the primary reason expressed for the resurrection of the Aztlán myth by Chican@ nationalists.

Basu has reproduced a table from Cohen's book (26; qtd in Basu 336-7) which delineates characteristics of a human diaspora. I reproduce it here both to buttress Basu's immediate claims and to complicate other claims to be made below:

Common Features of a Diaspora:

1) Dispersal from an original homeland, often traumatically, to two or more foreign regions.
2) Alternatively, the expansion from a homeland in search of work, in pursuit of trade or to further colonial ambitions.
3) A collective memory and myth about the homeland, including its location, history and achievements.
4) An idealization of the putative ancestral home and a collective commitment to its maintenance, restoration, safety and prosperity, even to its creation.
5) The development of a return movement that gains collective approbation.
6) A strong ethnic group consciousness sustained over a long time and based on a sense of distinctiveness, a common history, and the belief in a common fate.
7) A troubled relationship with host societies, suggesting a lack of acceptance at the least or the possibility that another calamity might befall the group.
8) A sense of empathy and solidarity with co-ethnic members in other countries of settlement; and
9) The possibility of a distinctive creative enriching life in host countries with a tolerance for pluralism.

Items 3-6 and 8 are often, if not always, accomplished via exchange of origin and journey stories. For both Basu and Cohen, a mythologized homeland narrative arises from diaspora, as if the dispersal creates a need that the narrative is created to fulfill. And this happens in most or even all of the diasporic cases studied by the two authors.

What emerges here is a concept of homeland narratives as multi-function tools, operating with equal efficiency for proprietor, as for voluntary and involuntary expatriates. Above I have posited that these stories compensate for various insults to territorial instincts, us/them dichotomies, and involuntary movements. In turn, the idea that a narrativized "homeland" is *necessary* to personal and cultural identity for both indigene and expatriate accretes to these latter categories as empirical cases are examined. Examples of cultures that fit Cohen's table of characteristics, but that aren't quite diasporic, suggest that journey/identity narratives with mythologized homelands like those described for the Scottish serve a broader spectrum of functions. Basu, together with his respondents and supporting theorists, all assert, while demonstrating via ethnography and wide-reaching analysis of journey and origin narratives, that people who leave their birth-country construct mythologized narratives about their homeland to establish and maintain their personal and cultural identities. There is considerable evidence that this homeland-construction is more than just a palliative for unhappy émigrés.

Aztlán and other symbols of the Chican@ nationalist movement arose out of the grape-worker strikes in California in 1965, led by Cesar Chávez. The word "Aztec" is a misnomer, for people who are more accurately referred to as Náhuatl. The word "Aztlán," meaning, ironically, "whiteness" (Leal 9, is taken from Náhuatl myths about the Aztec homeland, which is described as an edenic place with distinctive topography and an abundance of natural resources, where the gods live, and men never grow old (Leal 9-10). Aztlán was first mentioned specifically in relationship to the Chican@ nationalist movement in a paper, "*El Plan Espiritual de Aztlán*," presented at a youth conference in Denver in 1969. That document states seven organizational goals and offers a six-step plan of action for the unification and liberation of the "mestizo nation," including the re-acquisition of former property, along with the eviction of the more recent occupants (Anaya 1-5).

For Chican@s, Aztlán physically symbolizes those portions of the southwestern United States which Mexico gave up in the 1848 Treaty of

Guadalupe Hidalgo, when a considerable number of natives of Mexico suddenly became "Americans," though with scant say in the matter. The ancestral homeland of the Náhuatl people is vaguely posited to exist somewhere in these annexed lands. Likewise, tenuous connections to the modern and historical Pueblo people are alluded to in modern writings about Aztlán, which also serves other symbolic functions.

The Hispanic/Latin@ people of the U.S./Mexico border face particular identity challenges. First, their chaotic genetic origins are historically documented—interbreeding (frequently via enslavement and rape) between Native Americans and invading Spanish and other Europeans produced many of the people who now self-identify as Chican@. They thus originate from disparate places and cultures, whose contact was for the most part highly conflicted, and whose history and behavior are very difficult to mythologize. Second, they live in lands that were appropriated by other Europeans, lands that were taken away from their ancestors, usually by force. Third, the history and even the prehistory of the Southwestern United States are increasingly well-documented, closing gaps that might otherwise be filled by imagination and desire. Other complications will be discussed below. But the deployment of Aztlán as a cultural homeland is complicated by these contrary historical facts.

In their introduction to *Aztlán: Essays on the Chicano Homeland*, editors Rudolfo Anaya and Francisco Lomeli capture much of what has been said above about origin tales:

> The concept of homeland occupies a central position in the thought and development of most cultures. In some, as in Native American cultures of the Southwest, the place of origin is crucial to spiritual orientation. For other groups of people, such as the Hispanic community of this country, defining the origins of a homeland was a necessity for survival ... Aztlán brought together a culture which had been somewhat disjointed and dispersed,

> allowing it, for the first time, a framework within which to understand itself. (ii)

At the same time they are conscious of the tenuous nature and imaginary quality of the posited homeland, employing language that foregrounds the fictive and conceptual nature of the construct: "... Chicanos *reappropriated* Aztlán as their homeland ... the *concept* of Aztlán ... to define a *mythic past and history* ... a collective *symbol* ... its *mythic* nature ... *symbolic* nature ... the *legend* of Aztlán ...exists at the level of *symbol* and *archetype* ...(ii-iii; emphases added).

And they take advantage of this imaginary quality, to acknowledge it, accept it, celebrate it, and turn the nebulous foundation of the concept to their advantage:

> For some, Aztlán may appear to be an anachronism from the pages of Aztec or Náhuatl mythology that was lost in the annals of Native American history. The revival of it by Hispanicized mestizos attests to the role of myth in a culture where facts do not outweigh faith, beliefs, or the power of oral tradition ... [Aztlán] is a symbol which speaks of origins and ancestors, and it is a symbol of what we imagine ourselves to be. It embodies a human perspective of time and place. (iii)

Thus, the Aztlán homeland is not dependant on potentially contestable claims of truth and ownership. Its role is consciously defined as symbolic, in terms that Basu, Cohen, Caftanzaglou, and other theorists would find familiar. Unlike most origin myths and journey tales, here the tale-tellers are conscious of and acknowledge the artifice. In addition, the Aztlán myth was used by the Chican@ nationalist movement: to claim dispersal from a common homeland; to assert a collective memory and myth about that homeland; to promote an idealization of that homeland and a collective commitment to its maintenance; to emphasize a symbolic return to the edenic state of unity; to attempt to build and sustain a strong ethnic consciousness. The movement has a troubled relationship with the host society, emphasizing in recent times a solidarity with Hispanic/Latin@ people in other countries. In short, it has most of the characteristics of a diaspora tabulated by Cohen and Basu above. Still, the Chican@ case is not quite one of diaspora. In simple terms, the people of northern Mexico and the Southwestern United States did not leave their territory—it was taken away from them. But the appropriation of Aztlán here has been, at least in part, an appropriation of a diasporic identity, with the apparent function of unifying a people whose dispersal is not the result of traditional migration. This is, then, an attempt to deploy an appropriated origin myth by a culture which stands somewhat outside the traditional definition of diaspora.

Cohen and Basu's last diasporic characteristic, "The possibility of a distinctive creative enriching life in host countries with a tolerance for pluralism," has been ironically troped. The Aztlán myth, in the first place an origin story of the Aztec people, then appropriated by twentieth-century Chican@ nationalists for their own purposes of unification, has been re-appropriated by anti-immigration forces in the United States and turned to a different function. Well-known American political conservative and one-time presidential candidate, Pat Buchanan, has re-conceived the Aztlán myth in his recent book *State of Emergency: The Third World Invasion and Conquest of America*. In chapters entitled "A Grudge Against the Gringo" and "The Aztlan Plot," Buchanan takes advantage of the Aztlán homeland's mythical origins and fictive ambiguities to denounce it as an invention concocted for purposes of "racial" invasion. Buchanan claims that at the time of annexation, the lands that became what is now the Southwestern United States contained only a few thousand Mexican nationals, but now contain several million, most of whom have entered in the last few decades (108). In these terms the modern Chican@ movement does qualify as a diaspora—a migration away from poverty and racism in Mexico toward the wealth and stability of the United States. For Buchanan, the diasporic tale is not about a prehistoric journey from a mythical and ambiguous Aztlán, but embodies a real, modern journey, one of emigration, of conquest, of *reconquista* [reconquest] from south of the Rio Bravo north to the United States (93-137). Buchanan decries the Aztlán plan as a literal invasion, reviling it with rhetoric that emphasizes its threats to U. S. assets, territorial boundaries, and supposed cultural identity:

> This, then, is the Aztlan strategy: endless migration from Mexico north, the Hispanicization of the American Southwest, and dual citizenship for all Mexican-Americans. The goals: Erase the border. Grow the influence, through Mexican-Americans, over how America disposes of her wealth and power. Gradually circumscribe the sovereignty of the United States. Lastly, economic and political merger of the nations in a binational union. And in the nuptial agreement, a commitment to share the wealth and power.
>
> Stated bluntly, the Aztlan strategy entails the end of the United States as a sovereign, self-sufficient, independent republic, the passing away of the American nation. They are coming to conquer us. (128)

In her study *The Myth of Matriarchal Prehistory: Why an Invented Past Won't Give Women a Future*, Cynthia Eller, a professor of religion and women's studies, warns against the hazards of positing mythological past to provide a charter for modern behavior. She argues that the effectiveness of prehistoric tales depends on their truth content (91) and that progressive modern social movements do not require a past exemplar to be effective (180-188). Writing on the same topic, Joan Bamberger notes that the story that matriarchal rule appears to have been concocted by men to justify subsequent patriarchy—women ruled, and ruled badly, so they cannot be allowed to rule again (A pity this rule doesn't apply to modern political parties!) (Bamberger). In turn, Michelle Scalise-Sugiyama, a literary Darwinist, has written a convincing argument that the versions of the Oedipus myth found in various cultures arose and were repeated to warn against incest and patricide (Scalise-Sugiyama).

The point in both cases is that prehistoric myth does not describe a factual past that can be projected into the here and now, rather, myth is as much a control mechanism as it is a liberating tool. Mythological tales can cut both ways—one significant element in the Aztec conquest was the pervasive myth that an Aztec god with pale skin and a red beard, would return to overthrow the then-current Náhuatl rulers and take control of the region. The myth predicted that this Quetzalcoatl would reappear in about the time and place that Cortes and his company of would-be *conquistadores* landed on the Mexican coast. This myth undercut Náhuatl resistance to the invaders and hastened the Conquest. Again, myth deployment can be risky business.

Buchanan's revision, like the Aztlán myth it re-co-opts, contains significant mythologizing and ambiguity in its own right, demonstrating some of the hazards of fictionalized origin myths. At least some of the power of these origin stories comes from their semblance as "truth." They are deployed, at least in part, to assert land rights. The first paragraph of *El Plan Espiritual de Aztlán* specifically mentions "reclaiming" the birth land of "our forefathers" (Anaya 1). Chican@ nationalists assert an Aztec origin in the American Southwest, and thus claim that that part of the United States belongs to them, because the Aztecs were there first—a claim with no historical substantiation. In more recent times, land ownership has become a matter of law, rather than tradition, and debates about factual claims now dominate it. Imaginative origin myths are at risk in this new world.

Beyond this configuration, both the Aztlán myth and Aztec historical reality offer a dubious factual basis for any sort of progressive social movement. Again, all evidence indicates that most Native Americans came into the Americas from Siberia in various waves and expansions, following environmental fluctuations and food resources, moving away from population pressures, invading, emigrating, displacing. Less than 500 years (circa 1050 AD to 1525 AD) of Náhuatl history is "known," however tenuously, leaving as much 14,000 years of journey and settlement veiled in mystery. Some Chican@ nationalists are trying to claim ownership of the desert Southwest via affiliation with the Aztec people, over against a highly detailed archeological record that shows that the Aztec were far from the first occupants of either central Mexico or any hypothesized Aztlán homeland. Instead, they were brutal, relatively modern invaders themselves, overrunning and enslaving previous "indigenous" populations who had occupied North and Central America for over 10,000 years. No historical narrative of a Náhuatl homeland and diaspora would be complete without mention of the multiple homelands, migrations, and displacements that trace the route back through Beringia. To select a particular "homeland" or diasporic motive for this hegira would be cherry-picking, at best. For instance, no one in the Chican@ nationalist movement has advocated a *reconquista* of either Mexico City or Siberia, the first and last "homeland" of the Náhuatl migration to the Americas. While Spanish conduct in both Old (the Spanish Inquisition) and New (the Mesoamerican conquest) Worlds would sicken any humane modern observer, the Aztec domain—functioning as one of the most sophisticated and successful of all city-states, east or west—also ranks as one of the most exploitive, bloodthirsty, and sexist empires in all human history. European reports, Náhuatl ethnographies, Aztec codices, and the stone temple carvings that remain, record a pervasive history of the Náhuatl rulers' torture of ingroup and outgroup members alike. Part of the reason for the downfall of the Náhuatl was that their Native American neighbors readily assisted the Spanish invaders, for they were willing to do anything to stop the forced tithing, enslavement, and outright butchery that accompanied Náhuatl rule (de Sahagún; Levy).

The people the Aztlán myth is intended to unify are descendants not just of Spanish conquerors and Aztec conquered, but also of other European and Native American groups, French and Karankawa along the Gulf coast, German and various native tribes in central Texas, Spanish and Hopi and Navajo in the Four Corners region, and African-American and Native Americans across the

South, along with Kiowa, Apache, Tonkawa, and other tribes that historically have populated the U.S./Mexico border. At the same time the Chican@ ethnic label typically excludes other progeny of European and Native American couplings across the northern regions of North America, offspring of Scandinavian, French, English and Eastern Europeans mingling with Ojibwe, Sioux, Iroquois, and other Native American groups who don't quite fit under the Aztlán umbrella. The factual claims, implicit or literal, embedded in the Chican@ nationalist movement: that modern Latin@s are descendants of the Náhuatl and the Spanish; that they originated somewhere in the American Southwest; that the Aztec state was an historical model worth emulating, are highly selective, essentializing, generalizing, and clearly problematic. From a factual standpoint, the Aztlán myth would seem to provide a poor foundation for a modern movement bent on social unification.

A host of earlier critics familiar with the Chican@ movement (Chavez 1984; Anzaldua 1987; Alarcón 1992; Perez-Torres 1995) have critiqued the deployment of the Aztlán story, with American Studies researcher Lee Bebout denouncing both the Chican@ appropriation and Buchanan's re-appropriation, arguing in his dissertation and a subsequent conference paper that both versions of the story "erase the material conditions and lived-experiences of the immigrants invoked in articulations of a "Bronze Brotherhood" or a "brown invasion." (Bebout).

Here then an "Aztlán" homeland myth, probably functioning for the Náhuatl in the identity-and-cohesion function theorized by Basu and others above, is appropriated by a second, partly derivative culture to serve similar functions, then is co-opted by a third culture to counter the functions for the second and, by several critical claims, ultimately erases the material reality of those it was borrowed to help. This is a bewildering cycle of functions, but at the same time it is evidence of the influence of journey and origin tales on cultural identity construction.

Yet most theorists seem to agree that all origin myths contain at least a degree of the fictive, and that their identity functions are not dependant on their "truth." Many of the modern deployers of the Aztlán myth acknowledge, and in fact even celebrate, this independence from verity. Origin myths help provide identity and cohesion for human individuals and cultures across the planet. As a modern instantiation, the Aztlán myth of the 1960s and 1970s, although consciously disengaged from the illusion of truth that underlies other origin tales, both embodies and exemplifies what modern theorists are claiming for origin tales in other cultures.

The Chican@ nationalist movement originated in the late 1960s, and thus has had a bare half-century to evolve. But by most lights it seems to have had some success. A subsequent florescence in arts and letters has been accompanied by an influx of "Indohispanics" (Anaya's term) into academia and politics, together with a rise in population percentages of Hispanics across the United States. Time will be the ultimate judge, but this self-conscious exercise in narrative identity building appears so far to be working.

Works Cited

Alarcón, Daniel Cooper. "The Aztec Palimpsest: Towards a New Understanding ofAztlán, Cultural Identity, and History," in *Aztlan: A Journal of Chicano Studies* 19:2 (1988-90): 33-69.

Anaya, Rudolfo A., and Francisco Lomeli, Eds. *Aztlán: Essays on the Chicano Homeland.* Albuquerque: University of New Mexico Press, 1989.

Anzaldúa, Gloria. *Borderlands/La Frontera: The New Mestiza.* San Francisco: AuntLute, 1987.

Basu, Paul."Hunting Down Homeland: Reflections on Homeland and the Search for Identity in the Scottish Diaspora," in *Contested Landscapes: Movement, Exile, and Place.* Eds. Barbara Bender and Margot Winer. Oxford: Oxford UP, 2001: 333-348.

Bamberger, Joan. "The Myth of Matriarchy," in *Woman, Culture, and Society.* Eds.

Michelle Zimbalist Rosaldo and Louise Lamphere. Stanford; Stanford UP, 1974: 263-280.

Bebout, Lee. "Returning to Aztlán: Chicano Nationalism, the Nativist Right, and the Immigration Debate." Paper presented at the annual meeting of The

American Studies Association, Philadelphia Marriott Downtown, Philadelphia, PA, Oct 11,2007. 2008-06-25 http://www.allacademic.com/meta/p186472_index.html

Bender, Barbara, and Margot Winer, Eds. *Contested Landscapes: Movement, Exile, and Place. Op. cit.*

Bowlby, John. "Grief and Mourning in Infancy." *The Psychoanalytic Study of the Child* 15, 3-39. 1960.

Brown, Donald. *Human Universals*. Philadelphia: Temple UP, 1991.

Buchanan, Patrick J. *State of Emergency: The Third World Invasion and Conquest of America.* New York: Macmillan, 2006.Buller, David J. *Adapting Minds: Evolutionary Psychology and the Persistent Quest for Human Nature*. Cambridge: MIT Press, 2005.

Caftanzoglou, Roxanne. "The Shadow of the Sacred Rock: Contrasting Discourses of Place under the Acropolis," in *Contested Landscapes: Movement, Exile, and Place. Op. cit.*: 21-35.

Casimir, Michael J. "The Dimensions of Territoriality: an Introduction." *Mobility and Territoriality.* Ed. Michael J. Casimir and Aparna Rao. New York: Berg, 1992.

Casimir, Michael J., and Aparna Rao, Eds. *Mobility and Territoriality. Op. cit.*

Chávez, John R. *The Lost Land: The Chicano Image of the Southwest.* Albuquerque: U of New Mexico P, 1984.

Cohen, Robin. *Global Diasporas: An Introduction.* London: UCL Press, 1997.

Eibl-Eibesfeldt, Irenaus. *Human Ethology*. New York: Aldine De Gruyter, 1989.

Jobling, Ian. "The Psychological Foundations of the Hero-Ogre Story: A Cross-cultural Study." *Human Nature*, 2001.

Kiberd, Declan. *Inventing Ireland.* Cambridge: Harvard UP, 1995.

Leal, Luis. "In Search of Aztlán," in *Aztlán: Essays on the Chicano Homeland. Op. cit:*7-13.

Levy, Buddy. *Conquistador*. New York: Bantam, 2008.

Marshall, John, and Claire Ritchie. *Where are the Ju/Wasi of Nyae Nyae?* Communications No. 9. Cape Town, South Africa: Center for African Studies, University of Cape Town, 1984.

Parkinson, Brian, and Andrew M. Colman, eds. *Emotion and Motivation.* New York:Longman, 1995.

Pérez-Torres, Rafael. "Refiguring Aztlán." *Aztlán: A Journal of Chicano Studies.* 22:2(1997): 15-41.

Rapport, N., and Dawson, A., eds. *Migrants of Identity: Perceptions of Home in a World of Movement.* Oxford: Berg, 1998.

Sahagún, Bernardino de. *Florentine Codex: General History of the Things of New Spain.* Eds. Charles Dibble and A. J. O. Anderson. Salt Lake City: U Utah P, 1982.

Scalise-Sugiyama, Michelle. "New Science, Old Myth: An Evolutionary Critique of the Oedipal Paradigm." *Mosaic* 34:1 (March 2001): 121-36.

Steward, Julian H. "Economic and Social Basis of Primitive Bands." *Essays in Anthropology Presented to A. L. Kroeber.* Berkeley: U California P, 1936. n.p.

Thonhoff, Robert. *El Fuerte del Cíbolo: Sentinel of the Bexar-La Bahia Ranches.* Austin: Eakin Press, 1992.

Wrangham, Richard, and Dale Peterson. *Demonic Males: Apes and the Origins ofHuman Violence.* Boston: Houghton Mifflin, 1996.

CHAPTER 11

Empathy and Conflicting Moral Norms: A Bio-Cultural Affective Analysis of *Babel*

The purpose of this chapter is to suggest how biology and culture work together, both to enable and impede different groups of viewers' ability to empathize with politically disenfranchised characters' perspectives in *Babel* (2006). Like other films in the transnational cinematic genre, *Babel* seeks to elicit and extend viewers' empathy through dramatic representations of characters "caught in the cracks of globalization" (Ezra and Rowden 7). The potential for such empathy appears to be built into the film's "affective design structure," to use Ed S. Tan's apt phrase. The thematic interconnections between the film's four plots, the care with which it provides detailed contexts for characters' emotional reactions, and the director's stated intentions to make the movie "about the point of view of others" (Iñárritu), all suggest that the film has been "emotively prefigured" (N. Carroll 68) to elicit and extend empathy.

Nonetheless, the perils that befall characters in *Babel* are not guaranteed to evoke viewers' empathy. In the months preceding the film's theatrical release in the U.S., the social and political climate were particularly fraught with warring emotions on border control, illegal immigration and terrorism. The film seeks to elicit an empathic understanding of situations related in varying degrees to these issues. *Babel's* attempt to motivate empathy for a Mexican woman, illegally employed as a nanny and subsequently deported, and for a Moroccan family who gets caught up accidentally in a politically charged manhunt for suspected terrorists, is mediated profoundly by the different moral compasses that various groups of viewers use to judge such contentious topics. Thus, while *Babel* tries to extend viewers' empathy, it will not always do so successfully. In this chapter I show how empathic understanding, although a universally shared affective and cognitive process, is mediated by moral considerations such as in-group loyalty and concerns for social purity. I discuss research in moral psychology suggesting that specific moral emotions used to respond to the film are concentrated in, and to some degree exclusive to, groups of viewers usefully labeled as "liberal" and "conservative." What I hope to make clear is that no matter how well the filmmakers have designed *Babel* to encourage their empathy, some viewers will likely find its emotional appeals melodramatic and unjustified. In some cases, viewers will empathize with the emotional states and actions of characters for whom the filmmakers intended no such response.

Defining or otherwise debating just what "empathy" is, where it comes from, and what it causes human beings to do, traces a long and varied history spanning disciplines across the humanities, the social sciences, and ethological biology. To further complicate matters, there is no one widely accepted definition of empathy. Moreover, diverse terms such as *sympathy*, *compassion* and *identification* are commonly thrown into the mix. My purpose here is not to adjudicate these debates. It is sufficient to explain briefly how empathy is understood by many psychological researchers, and also to articulate why this conceptualization is particularly useful for discussing how film viewers may or may not empathize with the plight of politically disenfranchised characters in *Babel*.

Empathy is a complex cognitive and affective process that involves more than simply "taking the perspective" of another person (Haidt "Moral" 862; Sober and Wilson 233). There is a general consensus among psychological researchers that "empathy involves not only the affective experience of the other person's actual or inferred emotional state, but also some minimal recognition and understanding of another's emotional state" (Decety 248). This explanation of empathy should not be controversial, as it encompasses both the affective and cognitive or "imaginative" component that is regularly emphasized by other psychological researchers (Hauser 190; Sober and Wilson 233; Zillmann 145-46) and film theorists (N. Carroll 64-67; Grodal 85-94; Neill 191). Empathy involves matching the emotional state of another; it also requires forming some belief about why that person is in that emotion state.

Empathy occurs in two primary ways. The first involves the "top down" "imaginative transposing of oneself into the thinking and feeling of another" (Decety and Jackson 54). This process occurs when we feel what a character in a film feels, which involves "viewer activation of affects and emotions in identification with the interests of a fictive being" (Grodal 93). The second way that empathy manifests itself is the "bottom up" recognition and mimicry of another's emotional expressions (Decety and Jackson, 54). This would include smiling along with a character when he/she receives good news or sharing a look of surprise when a character discovers something shocking. This sort of reaction is especially pertinent to filmic analysis, not necessarily because viewers must always mimic an emotion to empathize, but because characters' facial expressions, in general, convey a wealth of meaning that helps viewers become more certain about those individuals specific emotional states (Hoffner and Cantor; Tan).

The non-verbal communicative potential of universally recognized facial emotions bears directly on claims that have been made about *Babel* and its ability to communicate with viewers. One critic has noted that "our fruition" of the film is "in fact made more difficult because the narrative keeps changing its language settings," while "the characters' true intentions, their needs or desires" are necessarily "twisted in the process of communication" (Bertacco 139). The overt allusion of the film's title and the fact that six languages are spoken during the course of the stories it narrates does seem to suggest a thematic concern with communication failure, though people's receptivity to the emotional valence of facial expressions may facilitate the comprehension of characters' intentions, needs, or desires at the very point where language breaks down.

Babel has many scenes, some as long as eight minutes, where no dialogue is spoken. Nevertheless, these episodes convey important emotional information related to character development. For instance, late in the film, the deaf-mute teenage character, Chieko, meets a group of friends and spends the evening taking drugs, hanging out, and dancing. During the scene almost no dialogue is spoken. However, through context and facial expressions, viewers are able to see, literally, that Chieko transitions from being shy, to elated, then jealous, angry, and eventually quite sad when the boy she is interested in kisses her friend. After leaving the dance club where she witnesses the kiss, she walks the streets, her sadness and anger both evident and arresting on her face.

Additionally, even though all six languages used in the film—English, Spanish, French, Arabic, Japanese, and Sign Language—are subtitled in English, at least a dozen exchanges between characters are not translated. There is reason to suspect that these lapses are part of the artistic design of the film. Director Alejandro González Iñárritu has noted that "[t]he visual language" of cinema "is the universal language…It takes audiences, without words or translations, into places they could never reach in reality" (qtd. in O'Keefe 38). While language is central to filmic, as human communication in general, seeing, and subsequently feeling, that others are susceptible to the same pains, joys, and surprises as we are, helps motivate empathy, which can lead to greater sympathy and compassion with characters that represent groups of people whose perspectives and contexts we are seldom otherwise afforded.

While viewers may more easily empathize with a young girl who is upset by her friend stealing her love interest, other plots in *Babel* are much more politically contentious, such that empathic understanding is not necessarily guaranteed. Viewers may recognize that a character's facial expression is sad, angry, or joyous, but their willingness to match these emotions depends on their beliefs they have about why the character feels this way. As mentioned, two of the four plots in *Babel* focus on types of individuals who often are presented as unworthy of empathy, at least in some areas of American political and social discourse. One is Amelia, a middle-aged Mexican woman who works illegally as a nanny in San Diego. The others are Yussef and Ahmed, who are Arabic-speaking Moroccan boys whose accidental shooting of an American tourist precipitates a global manhunt for the "terrorist" suspects. Critically, Amelia, Yussef, and Ahmed are not simply characters to viewers of the film; they represent groups of people in the world that the average moviegoer may know little or nothing about. "Social schemas" such as "terrorist" or "illegal immigrant" "are capable of evoking affect directly. When someone fits into an existing person schema, then not only the meanings that are inherent in the schema are invoked but also the affect that has been attached to the schema by previous experience" (Tan 168). Once established, differential affective responses to culturally specified out-groups can become "mechanical…in the sense that the individual need not engage in explicit judgmental deliberations concerning the circumstances" (Zillmann 153) of the out-group member's behavior or situation. Thus, viewers are not responding to the plight of politically disenfranchised characters *de novo*; no matter how tragic some characters' outcomes appear to be, viewers bring emotionally charged knowledge and beliefs to the their understanding of the social schemas such individuals represent, as to the contexts that precipitate their emotional reactions.

The year *Babel* was released (2006) saw a dramatic increase in the public's awareness of immigration reform in the United States. During the spring preceding *Babel*'s October release in American theaters, dozens of pro-immigrant marches and protests were staged in major American cities (Shore). Some estimates suggest that over two million people rallied in support of immigrant rights, while these demonstrations were covered regularly by mainstream media. In addition to calling attention to efforts in the United States Senate to pass comprehensive immigration reform measures, many protestors' anger was directed at the House of Representatives' passing of The Border Protection, Anti-

Terrorism, and Illegal Immigration Control Act of 2005 (HR4437). Many observers felt that this bill effectively made felons of illegal immigrants, while severely punishing those who helped them ("Immigrant Rights"). HR4437 did not pass the Senate, though it did stoke viewers' emotional reactions to issues of immigration, border control, and terrorism in the months before *Babel*'s release.

It is not clear to what extent viewers of *Babel* were familiar with these pro-immigration protests. Likewise, we do not know if and to what extent viewers were influenced by either pro- or anti-immigration information. It is likely, however, as based on studies of viewers' empathic reactivity to news events dramatizing the plights of different groups, accord with their affective disposition toward those groups (Zillmann "Emotional"). That reactions such as empathy are mediated by affective dispositions means simply that how one feels emotionally about a group of people prior to hearing some news about them plays a significant role in whether or not one empathizes with their fortunes. What this means for empathy, according to narrative and media psychologist Dorf Zillmann, is that "empathic reactivity may be expected...wherever and whenever dispositions of liking exist. Dispositions of disliking, contempt, resentment, or open hatred, on the other hand, must be expected to foster counterempathic reactions. In such situations, the negative disposition virtually overrides empathic concerns, mostly because of moral considerations" (195). Importantly, research suggests that viewers' affective dispositions function "essentially the same for fictional and non-fictional representations," and "the empathic dynamics, in particular, are much the same" (203; see also Hoffner and Cantor 65). Thus, while *Babel* seeks to foster empathy for the illegal immigrant character of Amelia and the Moroccan family of Ahmed, Yussef, and Abdullah, how those who see the film feel and think about such groups and related political issues weighs significantly on their capacity to empathize.

As Zillmann mentioned, moral considerations play a significant role in mediating empathic response. Some recent research in evolutionary moral psychology suggests that the political groupings commonly referred to as liberal and conservative have rather specifically defined moral considerations that impinge directly on the nature of what qualifies as a moral concern. Moral psychologist Jonathan Haidt has constructed a model of moral behavior that suggests that liberals tend to limit their moral considerations to issues of harm and care, of fairness and reciprocity. On the other hand, conservatives' moral thinking, while including concerns for harm/care and fairness/reciprocity, is also informed by concerns for loyalty to the in-group, submissiveness before dominant individuals and authority, and a marked concern for bodily purity and social sanctity. Liberals decry that people who come to the United States illegally may be harmed in the process of crossing, abused while they work here, and are often cruelly deported. As such, the liberals' moral stance affirms that governmental policy is frequently unfair, even harmful, and so needs to be amended. Harm, care, and fairness trump all other concerns for individuals who organize their moral register in this way. However, if jobs are being denied to one's in-group on account of illegal immigrants, and if one believes that America has fundamental Anglo-Protestant values that should remain the core of the culture, illegal immigration, irrespective of the harm done to out-group members, will incite negative feelings toward groups who appear to be harming the in-group and diluting the purity of its identity (Huntington). Whereas most liberal people do not regard diluting the purity of a group's identity as a moral issue, "political and religious conservatives are more likely than political and religious liberals to moralize behaviors that do not involve direct harm" (Haidt and Graham 107). According to Haidt, what is lost on liberal viewers, accounting for many of the disputes between them and conservatives, "is that justice and related virtues (based on the fairness foundation) make up half of the moral world for liberals, while justice-related concerns make up only one fifth of the moral world for conservatives, who, in turn, have moral concerns that liberals simply do not recognize" as such (Haidt and Graham 99).

Haidt argues that these five categories, shared by liberals and conservatives to varying degrees, constitute "innate intuitions" that channel the development of specific and widespread morals and virtues across cultures. These "moral intuitions" are best thought of as similar to "taste buds," the sensitivities of which move from "crude, global judgments articulated using a small number of innate moral intuitions to highly sophisticated and differentiated perceptions, beliefs, emotional responses and judgments" (Hadit and Joseph). Every human being comes into the world with the potential to develop moral feelings for all five of the categories that Haidt and other researchers have identified. The moral categories are believed to stem from evolved predispositions to engage in parental care (harm), reciprocal altruism (fairness), tribal instincts (loyalty), to recognize status (submission), and to feel disgust (social purity/sanctity). The concentration of specific moral feelings in human groups is likely determined by a range of variables including geographical location, trade and immigration patterns, and specific reinforcement measures developed in each culture. Liberal values tend to be concentrated in port cities that have historically been nexuses for trade and immigration. Conservative values tend to be concentrated in tight-knit, discrete groups. The social dynamics and adaptive challenges for each environment are very different and may have pushed specific constellations of moral emotions to the forefront of each group's socio-emotional cosmology.

For both liberals and conservatives, narratives play an essential role in refining and reinforcing each group's moral "taste buds." By providing emotionally prefigured mental models for group members, liberals and conservatives recreate from generation to generation the appropriate way to interpret, understand, and empathize with the behavior of others based upon the specific moral concerns of the group. *Babel* is designed to do precisely this for liberal viewers: to elicit empathy and pro-social moral emotions like compassion and indignation by emphasizing the harm done to politically disenfranchised people around the world. The film is likely to be received well by liberal viewers because it has been emotively prefigured to appeal to their two principle moral sensibilities: harm/care and fairness/reciprocity.

The claim that conservatives have moral intuitions that liberals do not recognize is an important caution for liberal-leaning humanistic researchers. While it may *feel* right to conclude that Iñárritu ends a film with a "universalizing statement" that "ties all of the characters together in their shared humanity and mortality" (Newman 104), the way in which such universality is conveyed is pivotal to viewers' emotional recognition. The conclusion of *Babel* is a case in point. In the film's final scene, the deaf-mute character, Chieko, stands naked on the balcony of her apartment being comforted and consoled by her father. For some viewers this scene is an affirmation of the fragility of each individual, together with the need for care and tenderness in the wake of losing a loved one. While liberals and conservatives may both agree with this thesis concerning human needs, the way in which it is represented through the naked body of teenage girl may elicit negative moral reactions involving the sanctity and proper use of the body. If "the promise of a cognitively and affectively valuable final representation governs the emotional meaning structure" of a film (Tan 202), then the final moral-emotional representation in *Babel* may forestall empathy at the very point where the filmmakers intended it most. For instance, one response in *Christianity Today* recognizes that the scene is "meant to emphasize her desperation and vulnerability," though the reviewer is uneasy and feels that more should have been done to "preserve the actress's dignity" and that "these scenes may do more harm than good" (Overstreet).

Having established that *Babel*, though designed to elicit empathy, fights an uphill battle against the moral norms of different groups of viewers, I now want to provide a number of additional examples from the film that suggest how particular scenes can be read very differently, depending on the affective disposition of the viewer and the moral precepts that inform this disposition. I first provide a brief summary of the film for those who have not seen it.

Babel begins in Morocco as Hassan Ibrahim sells a rifle to his neighbor, Abdullah, who gives the rifle to his two teenage sons, Yussef and Ahmed, to protect the family's goats from jackals. In a moment of boredom and brotherly competition, Yussef and Ahmed dare each other to shoot at cars from atop a hill. Yussef shoots at a tour bus and strikes an American tourist in the neck. They run home and become consumed by guilt when they hear, wrongly, as it turns out, that the tourist has died.

The second narrative in the film follows the American tourist who is shot, Susan, and her husband Richard, as they are forced to seek emergency medical aide in the small desert town of Tazarine. Richard and Susan, we learn early on, have traveled to Morocco in an attempt to heal emotional wounds and unresolved resentment after the sudden death of their infant child. While waiting in Tazarine to be rescued by the United States government, Susan is cared for by a local doctor.

The third narrative focuses on Amelia, an illegal Mexican immigrant who is the nanny for Richard and Susan's two other young children, Mike and Debbie. Because Susan has been shot, Richard tells Amelia that she must continue looking after the children, even though this means she will miss her son's wedding back home in Mexico. Desperate to see her son married, Amelia takes Mike and Debbie to the wedding. Amelia, Mike and Debbie are driven to Mexico by Amelia's nephew, Santiago. After an altercation with the border patrol just before reentering the United States after the wedding, Santiago panics and flees from the police, abandoning Amelia, Debbie, and Mike in the desert at night. The border patrol eventually finds Amelia and the children later the next day, only after they nearly die of dehydration.

The fourth and final narrative in *Babel* focuses on Chieko, a deaf-mute Japanese teenager whose mother recently committed suicide. Unable to deal emotionally with her mother's death, and resistant to her father's attempts to connect with her, many of Chieko's scenes involve awkward and inappropriate attempts to gain sexual affection from men. Chieko's narrative ties in with the other plots of the film via her father. While big-game hunting in Morocco some years before, her father gave the gun used to shoot Susan to Hassan Ibrahim as a gift.

Ultimately, Susan and Richard are airlifted to a hospital, where we are led to believe that Susan will survive. After fleeing from the police, Yussef, Ahmed, and their father are spotted and cornered. Ahmed is shot and killed, and Yussef and his father are taken into custody. After being found with the children in the desert, Amelia is arrested and deported back to Mexico. Chieko and her father come to realize their shared pain and loneliness after the mother's death, and the film ends with Chieko naked and vulnerable, crying in her father's arms as he consoles her.

A common focus in transnational films like *Babel* is "the ways in which physical mobility across national borders necessarily entails significant emotional conflict and psychological adjustment" (Ezra and Rowden 7). As the summary indicates, this is the focus for the stories of Amelia, Richard and Susan. All of their fates are altered dramatically either by crossing a national border or visiting another country. Iñárritu has made this representational focus clear as well: "The film really talks about all of us on the other side of what the average American citizen has been able to experience from the TV screen without having crossed the border into the world out there. It seeks to reveal that other truth." Although Iñárritu states that he sought to construct all characters in the film with "dignity and compassion," "[t]he most important thing for me," he admits, "was not to tell a story from the point of view of tourists in some country. That's a very common thing, to portray another culture in the light of our eyes, of our reality…I tried hard to see…the drama of their world through their eyes." Critics agree, noting that "the sensibility of *Babel* is that of someone from a Third World country" (Bertacco 140). Given his project to emphasize how characters see the actions of tourists, many scenes in the film function to critique how western characters perceive and treat other cultures. More specifically, these scenes seek at once to confirm and to challenge specific aspects of conservatives' moral foundations such as in-group loyalty and concerns with purity.

The first scene in which we encounter Susan and Richard clearly attempts to depict Susan's moral reactions to being away from home in a foreign place. Fearful of getting sick in a local Moroccan restaurant, she throws their ice cubes into the dirt and then nervously rubs sanitizer on her hands, as if the glass she just touched were diseased. Susan's strong reaction to the food, water, and even physical objects in Morocco, bordering on disgust, is reinforced by the scene when she is treated for her gunshot by the locals in the small town of Tazarine. She is visibly frightened by the unfamiliar people and recoils from their attempts to help her. At one point she yells, "Richard, don't leave me!" Of course, one might contest that Susan's obsessive fear of others is not being critiqued by the film at this point; she has, after all, just been shot and is supposed to be in a state of panic and shock, which may explain why she does not want anyone to touch her. This explanation, however, is unlikely for two reasons. First, viewers are privy to a translation of the local doctor's assessment of her injuries. The doctor notes that the bullet has not touched her spine, but her clavicle is broken. The function of including this brief medical diagnosis is to convince viewers of the local Moroccan doctor's competence in western medicine, and that it is Susan, in fact, who is being unnecessarily fearful. Second, Susan's negative affective disposition toward foreigners is shown to have influenced her son's emotional reactions to crossing the border into Mexico in a prior scene.

When Mike is taken to Mexico to attend a wedding, he nervously says to Amelia, "My mom told me that Mexico is dangerous." Amelia's rambunctious nephew, Santiago, turns to Mike and says jokingly, "Yes, it is! It's full of Mexicans!" Mike's concern and Santiago's jest serve to link the fearful stereotypes that have originated with Susan and have been passed on to her son. As this scene continues and they cross the border, an upbeat song accompanies a visual medley of shots, and we see Mike looking intently out the window at the teeming life of Tijuana. Some viewers might see this scene as working to amend Mike's negative perceptions of Mexico. Through the car window he sees children playing in the streets and mothers taking children to dance practice. These are all slice-of-life images that suggest a shared human existence on both sides of the border. However, intermixed with these empathy-potentiating images are glimpses of prostitutes and pimps, barred windows and buildings, and flies swarming around a piece of meat left on a plate. While the medley presents a range of images common to both the U.S. and Mexico, the elements of the scene designed to elicit shared humanity might be overbalanced by elements that confirm the negative affective dispositions of some viewers toward Mexico and Mexicans.

The principle moral reaction that conservatives might feel during this scene, but that liberals might not, involves purity and sanctity. Conservatives' concern with purity has a double function related to disgust, a physical reaction evolved as a response to the source or "contact history" of food. On the other hand, disgust and its corresponding facial expressions have been co-opted during humans' evolutionary development to condemn social transgressions involving sexuality, drugs, and body medication (Haidt "Moral Mind" 857). Thus, the medley scene, while ostensibly created for viewers to perceive continuities between different ways of life on both sides of the U.S./Mexico border, may be registered very differently by conservative viewers who see images of fly-infested food and illicit sexual activity. Liberals may not necessary like such images, but they are not inclined to assess them along moral lines. If these images confirm conservative viewers' affective dispositions toward Mexicans, empathy for people of that origin might be withheld, and Mike's fearful preconception of Mexico as dangerous will be the emotion with which viewers empathize.

The Tijuana medley scene is visually and in some ways emotionally analogous to a later scene in which a group of European tourists stare out of their tour bus windows at the denizens of the desert town of Tazarine. The sight of burka-covered women and stolid faced men on rooftops elicits looks of concern from the Europeans. While the bus is parked, the tourists survey the locals, and a story is floated around concerning a bus-load of tourists whose throats were slit in some other small town in Egypt. Even though Susan's life hangs in the balance, their fear of the locals rises to such a pitch that the other tourists abandon the wounded woman and her husband.

The emotional unease with which Susan, her son, and the European tourists encounter other cultures could invite empathy from viewers who believe that third-world countries are dangerous and dirty. However, by providing viewers with more information than the characters have, the film invites us to pity, perhaps even to condemn them for their narrow-mindedness. Several reviews of the film develop this "Ugly American" critique. The "selfish westerners," as one reviewer put it, are an "indictment of how rich people on vacation behave while the poor villagers show wordless compassion" (Alexander). Another notes that "the tourists' paranoia—they are American, French, and British—is in no way justified by the treatment they receive in the village" (Chocano). To dissuade viewers from empathizing with the tourists' emotional reactions to foreign places, the film encourages viewers to see that Mexico, like America, is a place where family weddings and other celebrations occur. Likewise, Tazarine is home to a knowledgeable doctor who saves Susan's life. It is also the birthplace of the unnamed altruist who guides Richard and Susan there, refusing money for his kindnesses. As such, although the film provides points of emotional identification for conservative viewers, it never ceases questioning the emotional legitimacy of such identification by offering models of kindness and generosity to contrast them.

It is difficult, however, to say at what point exactly a viewer who empathizes with Susan's, Mike's, and the retreating tourists' fears will come to realize that such empathy is misplaced. While viewers' prior moral considerations will mediate how they respond, the emotional *intensity* of those prior beliefs matter (Zillmann 202). As with any trait, individuals vary in how fearful of others they are. Personality and individual experience probably make some viewers prime candidates for experiencing novel extensions of empathy and other pro-social moral emotions while watching the film. It is also possible that viewers' perception of how *deserving* the locals are for the treatment they receive determines whether or not they withdraw empathy from the tourists and extend it toward the locals (Zillmann 202). That kind of response appears in the reviews by Alexander and Chocano cited above. Conservative viewers, after all, are not incapable of having moral responses based on care/harm and fairness/reciprocity. It is just that other competing moral concerns attenuate a singular focus on care and justice that is so central to liberals' moral feelings. If viewers come to feel that the locals are not deserving of fear, they will be more likely to extend empathy in the manner the film has intended. For instance, when Richard discovers that no ambulance is available to take Susan to a hospital, he punches the door next to the local policeman's head and screams, "This is YOUR fucked up country! YOU do something about it!" His rage, which is precipitated by fear for wife's life, is understandable. However, as the local policeman silently accepts his humiliation, one is left with the feeling that he is completely undeserving of such treatment. Viewers may be moved to sympathize with him if they come to understand, as the film suggests, that Richard is acting as he does from his privileged position as an American.

Other scenes are designed to be critical of such privilege and the resulting displays of power that the film associates with the United States. Viewers are aware that the gunshot that initiates the movie occurs in a thoughtless moment of childish heedlessness, though other characters do not understand what happened, and the unfortunate response to the incident accounts for much of the frustration viewers feel about ensuing events in the film. The U.S. government believes that the shooting is a terrorist act, which motivates the Moroccan police to begin their search for the shooter. Once viewers come to see that the Moroccan police are brutal and uncompromising, they quickly realize that any attempt to understand the real cause of the shooting is doomed to fail. The shooting soon becomes an international incident, while all attempts to locate its origins on an inter-personal level quickly evaporate. However, because the film depicts Yussef's and Ahmed's point of view before, during, and after the shooting, it is possible to form an understanding of their situation and to empathize with them. As they experience events from the boys' vantage point, viewers grasp their racking guilt, their inability to eat and sleep, and the frantic worry they feel when they learn that the police are pursuing their father, who is entirely oblivious to the chain of events slowly encircling him. While all viewers are likely shocked by the casualness with which Yussuf and Ahmed shoot at the passing cars, empathy for their situation is dependent on how viewers may feel about the film's stance on the United States government's response to the shooting.

On more than one occasion a voiceover newsbyte version of the shooting is heard, claiming that the United States has labeled the shooting a terrorist act despite the Moroccan government's insistence that terrorists have been eradicated from their country. Political pressure from the United States forces the Moroccan police to take swift, uncompromising action. This results in the Moroccan police pursuing Abdullah with a shoot first, ask questions later mentality, appropriate perhaps for a calculating terrorist but not for an oblivious farmer. As a result of this disproportionate response, Ahmed dies. The United States, the film suggests, over-determines the geo-political significance of the event. Because the shooting is dealt with impersonally and from a distance, confusion, harm, and ultimate injustice ensue, as two scenes emphasize. After Ahmed is killed, Yussef breaks the rifle used to shoot Susan on a rock and confronts the police. He tells them in a gesture that is at once submissive and accusatory, that he is the one they are seeking. The police captain, who is presented throughout the film as shrewd and willing to use violence to achieve his ends, looks incredulously at the "terrorist" he has pursued. The captain realizes the personal and local, rather than global nature of the tragedy that has unfolded in his province. The official version of events is wrong, garbled by so many distant assumptions and impersonal recommendations. As if to reinforce his sense of amazement, as well as to stoke viewers' indignation, he takes off his sunglasses to look at Yussef in disbelief.

At this precise moment the film begins cutting rapidly from scene to scene to underscore the moral significance of the captain's realization that unnecessary harm has been done. Yussef is seen sitting next to a rock as his brother's body is taken away. The next scene cuts to a descending medical helicopter that has come to airlift Susan and Richard out of Tazarine. Both Susan and Ahmed are being taken away on stretchers, yet the emotions viewers are encouraged to feel for each scene are strikingly at odds: sadness and anger for Yussef and relief for Susan. By juxtaposing the disparate fates of Susan and Ahmed, the filmmaker prompts us to conclude that human fates are often inextricably tied up with varying socio-political forces that determine privilege, access, and differential degrees of respect for life.

While the film does not directly ask "for whom the bell tolls," one critic argues that "it is evident that some men's deaths are more significant than others, and indeed that America is the bell-ringer" (French). For some, the film appears to be only a commentary on contingency and the exacerbating effects of poor judgment. Although these are themes in the film, liberal viewers who are attentive to moral considerations of harm and fairness are more likely to see globalization, rather than chance or poor judgment, as responsible for misfortunes in the film. Attacking the inequities that stem from globalization is central to many critics' reviews: "[t]he global village depicted in *Babel* is a harsh, unfair place" (French); the film seeks to illuminate the "inequalities that keep the global economy afloat" (Bertacco 139). "*Babel's* Tower of Babel—the human folly that brings down God's wrath—is globalization" (Stevens). It is because the director and writer feel the appropriateness of such criticisms, that they inject strong irony into the scene at the hospital where an American official comments on the "happy ending" to Susan's and Richard's ordeal. While this is a true assessment of Richard and Susan's experience, and one that is common to Hollywood films, its juxtaposition with Ahmed's death frustrates the release typically felt when characters are delivered from dire situations at the end of a film. This emotional disparity elicits indignation over the unfairness of such wildly disparate fates.

Nonetheless, it is likely that for many of those who watch *Babel*, the good name of America in this instance has been harmed. The empathy that the film asks its public to extend to Yussef and Abdullah must come at the expense of seeing one's in-group in a negative light. An alternative understanding of the same events is that after Susan's shooting, the safety of American tourists was at stake, so the U.S. government acted swiftly and decisively to ensure that nothing else befell its citizens. Some viewers will see the American response as more egregiously harmful and unfair than was Yussef's and Ahmed's reckless target practice. Other people will feel that this conclusion makes an unfair generalization concerning an esteemed authority, one to which they feel a profound moral obligation and loyalty. The director has spoken to this concern, observing that "unfortunately, there is a certain type of American who thinks that the film is a criticism, when it's not. It is simple a commentary on …reality…[i]t is a very American sentiment, which interprets any kind of criticism as an attack" (Iñárritu). The "reality," of course, together with viewers' ability to empathize with how characters experience it, is informed by different viewers' moral concerns. Liberal viewers may understand that their conservative counterparts feel that the film attacks them, though they are not emotionally prepared to feel sympathy for such grievances. Actual harm to "real" people is what matters, so the suggestion that damage has been done to the perception of a group to which one should be loyal simply does not make sense. Many reviews of the film, although not cheerleading for U. S. foreign policy or for the value of in-group loyalty, seem put off by the film's suggestion that someone other than Ahmed,

Yussef, and their father is to blame for the events that unfold in Morocco. One reviewer is nettled by the film's "insultingly infantile stabs at geo-political commentary with Richard and Susan's story" (Rogers). Others see it as "a feel good movie for the politically correct" (Thompson), one that "leaves you feeling as if the movie's doing the weeping for you" (Burr). It would seem that some of *Babel*'s public, not having come to empathize or to sympathize with the Third World characters, are left with a clear, if not somewhat discomforting, message: "Is the lesson here that white people should steer clear of nonwhite (read: dangerous) countries?" (Holcomb), while "the subtext is 'run away!'" (Channel Four). Such cynical responses contrast sharply with the prototypical liberal response that "things tend to turn out a lot better in this world for people who are white and well-off" (Gillespie). Thus, while some viewers draw moralistically liberal and sweepingly ideological conclusions about harm done to characters in the film, others effectively negate these sorts of impressions, commenting on the melodramatic puerility of the director and the self-righteousness of his perceived audience. Similar moral perceptions of events are also in play with viewers' response to Amelia's story.

Amelia's narrative is representative of any number of personalized stories about the more than twelve million Mexican citizens living and working illegally in America today. Debates in the U.S. over immigration often reflect some of the worst aspects of our species' "us" versus "them" tribalism. As this paper endeavors to show, viewers' empathy with the hardships of such individuals is by no means guaranteed. Although viewers may empathize with Amelia, various American social discourses paint illegal immigrants as job-thieves who sneak into "our" country work to codify moral valuations that then directly impinge upon any emotional response to her plight.

In an effort to extend empathy to Amelia, the film seeks to highlight issues of fairness that are specific to illegal workers in America. For example, Amelia has to take Mike and Debbie to Mexico only because she cannot get time off from her "employers," Richard and Susan, who know that she has no official recourse against their demands that she miss her son's wedding. Initially, Iñárritu cleverly stokes liberal viewers' anger with Richard, but finally he redeems this character. Richard's phone call to Amelia telling her that she cannot go to the wedding occurs before we learn that it is Susan who has been shot. It would appear that "Richard's insensitivity has seemed to spark the series of events" that result in Amelia getting deported (McCormack 70). At the end of the film, after we learn that Richard implored Amelia to stay with the kids because of Susan's emergency, we are granted access to the emotional state that precipitated his demands. Viewers see Richard struggling to hold back tears as he talks to his son and pretends that nothing is wrong. Seeing his face at this moment leaves no room for argument that his demands were absolutely unjustified .

This scene invites viewers to reassess their assumptions about Richard's imperious behavior. Crucial for this effect is that this episode immediately follows the scene in which Amelia is deported. At the same time, such revelations so late in the movie do not mean that any initial assumptions that Richard is a heartless employer did not serve to prime liberal viewers' conclusions about who is responsible for harm and injustice in the film.

The second way the film seeks to encourage empathy with Amelia is by making the bulk of her story preceding deportation one of shared joys to which viewers can readily relate. Customs, skin colors, and languages may differ, but by emphasizing a wedding celebration as a cornerstone for cross-cultural empathy, the public is encouraged to feel that some of the most felicitous moments in our lives are shared universally. Viewers' understanding of such social schema as "illegal immigrant" is potentially shifted from job-stealing trespasser to devoted mother. By establishing empathy for Amelia early on, the filmmakers can use our emotional connection to elicit anger and indignation over her unjust treatment.

To accomplish this, the high merriment of the wedding is starkly contrasted by the high seriousness of the border crossing back into America.

Santiago's drunkenness, together with the border guard's demands for specialized paperwork that would allow Amelia to take Mike and Debbie out of the country, prompts Santiago to flee recklessly from the border crossing, abandoning his passengers in the desert. In many respects, the desert scene is one of the most emotional in the film, pitting Amelia's desire to stay and protect the young children against the need to leave them to find help more quickly. As much as we are likely to empathize with her fear, as much as we realize that Santiago is to blame for her predicament, here certain viewers may view Amelia contemptuously, rather than empathically. In fact, some may empathize more with the frustration and anger of Mike, who turns to her at one point and says through his tears, "You're a bad person." The irony of the boy's blunt summary of his nanny's character is that it arises from his childish logic. The film expects viewers to conclude that Mike is too young and morally immature to make an accurate assessment of Amelia's predicament. At the same time, some who see *Babel* will relate to Mike's fear and anger, which seems justified in light of the fact that Amelia "broke the rules" by taking him and his sister to Mexico. In-group loyalty plays a significant role in mediating those with whom viewers empathize. Liberals will likely focus on the injustice that forced Amelia to take the children to the wedding in the first place. They may also feel that the border-crossing guard's treatment of Santiago was unfair. Conservatives, on the other hand, are probably prone to believe that Amelia has broken the law in the country with which they identify. She is in the United States illegally and has endangered two American children after recklessly disobeying her employer. Empathy is a function of emotion matching, while forming some sort of minimal understanding, so which set of facts viewers utilize to build their understanding, as well as how they may have felt about immigration prior to their viewing this scene, will determine their willingness to extend empathy.

Describing the "affect structure" of any given film, Ed S. Tan suggests that "the intensity, nature, and dosage of each individual emotion over time and in relation to other emotions, has been preprogrammed by the filmmaker, even though such preprogramming is not always based on conscious reflection. There is, in any case, a plan behind the film narrative, and each microproposition of the text base has a certain role to play within this plan" (233). This chapter has argued that many scenes in *Babel,* no matter how carefully crafted to elicit and extend empathy for politically disenfranchised characters, are necessarily mediated by viewers' pre-figured emotional responses to the groups such characters represent. These affective dispositions, in turn, are grounded in and constrained by a finite number of moral considerations that are concentrated differently in groups of viewers commonly called liberals and conservatives, whose moral feelings are grounded in and constrained by human biology. These sentiments have, however, been developed differently in and disseminated through cultural systems designed to channel human social behavior toward adaptive outcomes. Neither group's moral intuitions are superior; they are, rather, competing socio-moral strategies.

Many liberal viewers will respond more or less predictably, as to how the filmmakers intended. This is because those who watch most non-mainstream and transnational films, and the people who review them, follow the same liberal moral compass as the director. *Babel* opened in seven major U.S. cities—Miami, Los Angeles, San Diego, Phoenix, San Antonio and Austin—all of which have significant Latino populations, and all of which have liberal-leaning newspapers that published sympathetic reviews of the film. At the same time, reviews appearing in *National Review* and *Christian Century,* both major conservative news outlets, labeled the film a "globalization melodrama" and dismissed its emotional appeal as cloying (Ross 54; Vineberg 51). Many other reviewers who work for more or less liberal media and who in times past have condemned conservative moral expressions in other films, saw *Babel* as pretentious, self-absorbed, and a glaring example of "self-conscious arthouse cinema that is custom-tailored and machine-tooled for the dinner-party demographic" (Bradshaw 9). The fact that even some professed liberals found the film's emotional appeal to be overdetermined indicates that caution and clarification are in order when shuffling viewers into two otherwise neat categories.

So, labeling groups of viewers as liberal and conservative in the face of individual variations may seem crudely reductionist and analytically unproductive. Political and religious affiliations, after all, are neither absolute nor all encompassing. Significant variation within and between such groups clearly exists. At the same time, if we want to explain how artists seek to persuade audiences to novel levels of cooperation and kindness in the age of globalization, we must consider how their attempts to extend empathy are designed to succeed, but also how they will inevitably fail, at least for certain viewers. As a starting point, what is required is a linkage of different kinds of viewers to different psychological dispositions, then discerning how such dispositions mediate the empathic process in question. The goal is not to erase variation by pretending there are only two monolithic types of viewers. Rather, it is to establish that two types of viewers' responses usefully account for subjectively discerned but theoretically guided patterns of variation between groups of viewers, while categories such as liberal and conservative may serve as parameters within which further variation may be studied.

Some readers may be underwhelmed by the thesis that morality mediates empathy and that scenes in the film are open to interpretation. Didn't we already know this before, without bringing the evolutionary moral psychology to bear? In some general sense, yes, this may have been understood intuitively as true. However, by linking filmic and literary analysis with concepts and findings in the evolutionary and cognitive sciences, we situate our interpretations and analyses within a network of mutually supportive hypotheses and findings about human behavior and culture. If an assumption about viewers' response is logically and evidentially consistent with theories about humans' evolved psychology, its development within cultures, and its relationship to narrative fiction, we have the right to be more confident that the phenomenon we are studying is actually—rather than only intuitively—true.

Moreover, bringing moral psychology to bear on viewers' response is necessary precisely because scenes in the film are not open to an infinite array of interpretations. The final scene of *Babel* where Chieko is comforted by her father may be interpreted in many ways. However, the range of these interpretations is finite, with some being more reasonable than others, given what we know about the filmmakers' intentions, viewers' psychology, and where this scene occurs in the film. Concepts and definitions from the evolutionary and cognitive sciences are useful here, because they tend to limit, rather than to extend indefinitely the range of interpretations. They provide us with an empirically defensible descriptive vocabulary for the type of moral reactions viewers commonly have to their social realities. There is no reason to assume that people somehow have a different moral psychology for processing film. These moral categories recur cross-culturally, enabled, as well as constrained by our species' evolutionary history. They do not represent surface level variations across cultures, but are probably based in human beings' evolved motivations to care, to enter into reciprocal exchange, to recognize status, to be loyal to a group, and to feel disgust. Establishing that such predispositions exist, not only in humankind, but in other animals, as well, means that the moral judgments arising from them can be generalized for the study of diverse audiences around the world. This understanding makes possible new insights from cross-cultural filmic and literary analysis. Optimally, hypotheses about how moral psychology mediates different groups' responses would be tested empirically on groups of viewers using films as elicitors of moral emotions. Such research would allow humanists to produce empirical, quantitative knowledge within their own disciplines, using the arts as a rich resource for the study of behavior and culture. In this way, the dialogue between the arts and the sciences becomes mutually beneficial (see Carroll *et al*).

In the wake of the cultural relativism that has dominated departments of literature and film studies over the last three decades, many people remain reluctant to embrace universal explanations for human behavior. Many intuitively feel that humans around the world share much in the emotional and cognitive realm, but they are resistant to the perceived implications of such continuity. This chapter has sought to allay such fears by showing that universality is most productively when seen as a starting point for examining variation within specific contexts. Cultural and socio-political analysis will always be central to any study of human behavior by nature of the type of cultural and political creatures we are. It is through empathy human beings enter into narratives: understanding how and what they come to feel about stories through their moral intuitions and cultural experiences can begin to shed light on the function of narratives for different groups, together with their role in motivating or impeding the human capacity to care. Because such capacities are grounded in biology, but developed and shaped in cultures, a bio-cultural framework for studying viewers' response demands our further attention.

Acknowledgements: I would like to thank Joseph Carroll and Leslie Heywood for their detailed comments on many versions of this chapter, as well as their encouragement and support.

Works Cited

Alexander, Victoria. "An Emotionally Visceral and Empathetic Triumph." Dec. 7 2006. www.filmsinreview.com. August 29, 2008. http://www.filmsinreview.com/2006/12/22/victoria%E2%80%99s-2006-christmas-reviews/.

Babel. Dir. Alejandro Gonzaléz Iñárritu. Written by Guillermo Arriaga. Perf. Rinko Kikuchi, Brad Pitt, Kate Blanchett, Adriana Barazza. Paramount Vantage, 2006.

"Babel Review." www.channel4.com. August 29 2008.http://www.channel4.com/film/reviews/film.jsp?id=159870&page=1.

Bertacco, Simona. "Language Settings and Self Expression." *Open Letter* 13:3 (2007): 132-149.

Burr, Ty. "Towering 'Babel' is an ambitious tale of human disconnection." Nov. 3 2006.

Boston.com. Aug. 29, 2008. http://www.boston.com/movies/display?display= movie&id=8820.

Carroll, Joseph, Jonathan Gottschall, John Johnson, and Daniel Kruger. *Graphing JaneAusten: Human Nature in Canonical British Novels of the Longer Nineteenth Century.* Forthcoming.

Carroll, Noël. *Engaging the Moving Image.* New Haven: Yale U P, 2003.

Chocano, Carina. "Babel." Oct. 27, 2006. www.calendarlive.com. Aug. 28, 2008.http://www.calendarlive.com/movies/reviews/cl-et-babel27oct27, 0,4536976.story.

Decety, Jean and Phillip Jackson. "A Social Neuroscience Perspective on Empathy." *Current Directions in Psychological Science* 30:2 (2006): 54-58.

---. "A Social Cognitive Neuroscience Model of Human Empathy," in *Social Neuroscience: Integrating Biological and Psychological Explanations of Social Behavior.*" Eds. Eddie Harmon-Jones and Piotr Winkielman. Guilford Press, 2007: 246-270.

Douthat, Ross. "And the Winner Is…" *National Review* 59:3 (March 2007): 54-55.

Ekman, Paul. *Emotions Revealed: Recognizing Faces and Feelings to Improve Communication and Emotional Life.* New York: Henry Holt, 2003.

Ezra, Elizabeth and Terry Rowden. "General Introduction: What is Transnational Cinema?" in *Transnational Cinema: The Film Reader.* Eds. Elizabeth Ezra and Terry Rowden. New York: Routledge, 2006. 1-12.

French, Phillip. "Babel." January 21 2007. www.Guardian.co.uk. August 29, 2008.http://www.guardian.co.uk/film/2007/jan/21/bradpitt.thriller.

Gallese, V. "Empathy, embodied simulation and mirroring mechanisms. Commentary on 'Towards a neuroscience of empathy' by Doug Watt." *Neuropsychoanalysis* 2:9 (2007): 146-151.

Gillespie, Eleanor Ringel. "Good Intentions Don't Always Make Good Movies." www.accessatlanta.com Aug. 29 2008. http://www.accessatlanta. com/movies/movies/etc/getCriticReview.jspd?criticReviewld=583.

Grodal, Torben. *Moving Pictures: A New Theory of Film Genres, Feelings, and Cognitions.* Oxford: Clarendon Press, 1997.

Haidt, Jonathan. "The Moral Emotions," in *Handbook of Affective Sciences.* Eds. R.J. Davidson and K.R. Scherer. Oxford: Oxford U P, 2003: 852-870.

---, and Craig Joseph. "The Moral Mind: How Five Sets of Innate Intuitions Guide the Development of Many Culture-Specific Virtues, and Perhaps Even Modules," in *The Innate Mind.* Eds. P. Carruthers, S. Laurence, and S. Stich. Oxford: Oxford U P, 2007: 3: 367-392.

---, and Jesse Graham. "When Morality Opposes Justice: Conservatives Have Moral Intuitions that Liberals May Not Recognize." *Social Justice Research* 20:1 (March 2007): 98-116

Hauser, Marc. *Moral Minds: How Nature Designed Our Universal Sense of Right and Wrong.* New York: Harper Collins, 2006.

Hoffner, Cynthia and Joanne Cantor. "Perceiving and Responding to Mass Media Characters," in *Responding to the Screen: Reception and Reaction Processes.* Eds. Jennings Bryant and Dolf Zillmann. Hillsdale, NJ: Lawrence Erlbaum, 1991: 63-101.

Holcomb, Mark. "Film Review: Babel." www.timeout.com. Oct 26-Nov 1, 2006. August 29 2008. http://www.timeout.com/newyork/articles/film/8939 /babel.

"Immigrant Rights Protests Rock the Country: Up to 2 Million Take to the Streets in the Largest Wave of Demonstrations in U.S. History." April 11, 2006. *Democracy Now: The War and Peace Report.* August 14, 2008. http://www.democracynow.org/2006/4/11/immigrant_rights_protests_rock _the_country.

Iñárritu, Alejandro Gonzaléz. Interview with Nathan Gardels. *NPQ: New Perspectives Quarterly* 24:2 (Spring 2007): 6-9.

McCormack, Karen. "Film Review." *Contemporary Justice Review* 11:1 (March 2008): 69-70.

Neill, Alex. "Empathy in (Film) Fiction," in *Post Theory: Reconstructing Film Studies.*Eds. David Bordwell and Noel Carroll. Madison: U Wisconsin P, 1999.175-194.

Newman, Michael Z. "Character and Complexity in American Independent Cinema: *21 Grams* and *Passion Fish.*" *Film Criticism* 31:1-2 (Fall-Winter 2006): 89-106.

O'Keefe, Alice. "Agent Provocateur." *New Statesman* (January 2007): 36-38

Overstreet, Jeffrey. "Babel." Nov. 3 2006. www.christianitytoday.com. Aug 29, 2008.http://www.christianitytoday.com/movies/reviews/2006/babel.html

Rogers, Nick. "Being diplomatic because of Inarritu and Arriaga's past success isn't worth it, given the irony that a story about communication breakdown so woefully fails to connect.Babel" is a bore." Nov. 10, 2006. www.rottentomatoes.com. Aug. 29 2008. http://www.rottentomatoes.com/m/babel/articles/1555413.

Shore, Elena. "Immigrant Rights Protests Spread: New Chapter in Civil Rights Movement." March 25, 2006. *New American Media, News Digest.* August 9, 2008. http://news.newamericamedia.org/news/view_article .html?articl e_id=70 da35086214d5eb9913e8d8114e0022

Sober, Elliot and David Sloan Wilson. *Unto Others: The Evolution and Psychology of Unselfish Behavior.* Cambridge, MA: Harvard U P, 1998.

Solórzano, Fernanda. "Babel." *Sight and Sound* 16:4 (2006): 16-17.

Stevens, Dana. "Coming Undone: Four Stories of Woe in *Babel.*" Oct. 26, 2006. www.slate.com. Aug. 29, 2008. http://www.slate.com/id/2152265/.

Tan, Ed S. *Emotion and the Structure of Narrative Film: Film as an Emotion Machine.* Trans. Barbara Fasting. Mahway, NJ: Lawrence Erlbaum, 1996.

Thompson, Luke. "Babel." Oct. 26, 2006. eonline.com. Aug 29, 2008. http://www.eonline.com/movies/e_reviews/index.jsp?movieuuid=37285962-f1a1-4b92-a593-307ab4244cf1

Vineberg, Steve. "Hours of Babel." *Christian Century* 123:25 (Dec. 2006): 51.

Watt, Douglas F. "Social Bonds and the Nature of Empathy." *Journal of Consciousness Studies* 12:8-10 (2005): 185-209.

Yacowar, Maurice. "Leaning Towards Babel." *Queens Quarterly* 114:1 (2007):122.

Zillmann, Dolf and Silvia Knoblach. "Emotional Reactions to Narratives About the Fortunes of Personae in the News Theater." *Poetics* 29 (2001): 189–206

Zillmann, Dolf. "Empathy: Affect from Bearing Witness to the Emotions of Others," in *Responding to the Screen: Reception and Reaction Processes. Op. cit.*: 135-167.

CHAPTER 12

Conceptual Knowledge of Natural Selection: A comparison study between Mexican-American and Anglo-American college students

> Learning is the goal of all instruction. Accurate assessment of learning is an important first step in determining the link between learning and teaching...Some disciplines, primarily physics and math, have made significant headway into unraveling the complex relationships between learning and teaching, often through the application of learning research.... (Libarkin and Anderson 394)

Unfortunately, this goal has yet to be achieved in the biological sciences in so far as natural selection (the core of the biological sciences) remains the single most misunderstood concept in evolutionary theory (McComas). Is the notion of natural selection challenging because students tend to interpret concepts and dynamic systems in static disjointed terms or is it because they identify a linear chain of causal forces to explain complex systems (Raia)? The theoretical framework used in this study supports the notion that scientific knowledge evolves when the learner moves from one paradigm to another in a nonlinear progression (Vosniadou). Conceptual change requires the learner to abandon his/her alternative framework and replace it with scientific concepts (Vosniadou).

In the quest to better understand students' conceptual knowledge, conceptual change, preconceived notions, beliefs, attitudes, etc., scholars continue to develop and refine assessment tools to gain better understanding of students' conceptual knowledge. Thus, the development and improvement of survey tools have established a point of reference in scientific research as scholars explore methods to ascertain the dynamic and complex relationships between learning and teaching (Anderson, Fisher, and Norman; Hestenes, Wells, and Swackhamer; Jensen and Finley, "Teaching Evolution"; Libarkin and Anderson; Yeo and

Zadnick; Zeilik, Schau, and Mattern). Disciplines like astronomy, biology, and physics have acquired and validated quantitative instruments that evaluate students' conceptual understanding (Anderson et al.; Libarkin and Anderson; Yeo and Zadnick; Zeilik et al). In addition, researchers have now started to explore and assess: (1) precise interrelationships among cognitive, affective, epistemological, and religious variables that lead to antievolutionary points of view among individuals who differ in age and educational backgrounds; (2) implementation and assessment of interventional tools that accurately promote cognitive models of evolution; and (3) methods reducing antievolutionary attitudes in students and teachers (Nehm and Schonfeld, "Measuring Knowledge").

Anderson et al. designed the Conceptual Inventory for Natural Selection (CINS) questionnaire to survey large numbers of students. CINS assess students' conceptual understanding of evolutionary theory (Anderson et al.; Libarkin and Anderson; Nehm and Schonfeld, "Measuring Knowledge"). Utilizing a modified version of the CINS, this study evaluates students' conceptual understanding or knowledge of *evolution* through natural selection after evolutionary theory instruction.

This study reports the findings on the student's conceptual understanding of evolutionary theory post instruction. The study assessed *conceptual knowledge* in terms of the breadth and depth of knowledge directly related to and represented by a specific domain of scientific principles and relationships among natural selection concepts_(Alao and Gunthrie). In addition, the study evaluates students' conceptual knowledge post instruction in traditional university teaching settings and compares two *ethnically different* student groups for differences in their level of understanding. Furthermore, this study identifies common misconceptions held by the students in general and by the two specific ethnic groups.

More specifically, this chapter compares conceptual knowledge of evolutionary theory through the process of natural selection in traditional pedagogy instruction of Latino and Non-Latino students attending five Hispanic Serving Institutions (HSIs) in Texas. No significant differences between the two groups regarding the most common misconceptions were found; however, noticeable differences were found when students were compared by religious affiliation. The theoretical implications of these findings are discussed below with regard to traditional pedagogy instruction and racial and ethnic identity.

Literature Review

Darwin's theory of evolution through natural selection fundamentally transformed the biological sciences. His evolutionary theory describes a process by which species come to obtain genetic characteristics or adaptations to the environment (Pidwirny). Evolution is driven by natural selection when organisms with favorable genetic characteristics have distinctive advantages to survive and reproduce at higher rates in a particular environment (Anderson; Balgopal; Kutschera and Niklas; Sadler; Stallings). "Because the nature of the environment is always changing, natural selection is continually influencing the genetic characteristics of the population through time" (Pidwirny ch. 9, sect. C). Hence, natural selection is the mechanism that accounts for the origin of new phenotypic variants or for the diversification of life over time as natural selection increases allele frequencies (Anderson; Kutschera and Niklas; Stallings). Because evolutionary theory unifies the biological processes which serve as the cornerstone for the life sciences (Anderson at al.; Balgopal; Sadler), the National Academic of Science and the National Research Council have challenged the biology teaching community to make evolution the center theme for biology courses (Anderson). If students have a comprehensive conceptual understanding of evolutionary theory, they will have the necessary knowledge to synthesize and integrate diverse biological concepts (Anderson; Bishop and Anderson, *Occasional Paper*; Demastes, Settlage, and Good) and achieve a better understanding of population change over time (Anderson; Bishop and Anderson, *Occasional Paper*; Demastes et al.).

Evolution through natural selection provides the theoretical framework for modern biological theory (Bishop and Anderson, *Occasional Paper*) and explains the diversification of all life: plants, animals, and microorganisms (Kutschera and Niklas). Therefore, science education is not only about making and measuring observations; but it is also about providing students with the knowledge and critical thinking skills to discover universal truths in order to protect our environment and improve the quality of life for humanity (Jensen). However, the relevant literature informs us that the evolutionary process of natural selection is misunderstood by many members of our society (Abd-El-Khalick and Lederman; Anderson et al.; Bishop and Anderson, "Student Conceptions"; Brumby; Clough and Wood-Robinson; Dagher and BouJaoude; Demastes, Good, and Peebles; ; Demastes et al., 1995; Lederman; Nehm and Reilly) and therefore, has a significant impact on how world situations are observed, addressed, and resolved (Alters and Nelson; Blackwell, Powell, and Dukes).

Prior to the 1970's, it was believed that incorrect responses revealed a student's lack of knowledge or lack of learning (Anderson et al.). However, studies conducted after the 1970's demonstrated that students' incorrect responses were not random at all, but followed predictable patterns (Anderson et al). The relevant literature continues to document students' difficulty in understanding key evolutionary terminology as well as comprehension of the biological processes associated with evolution and natural selection. Is it the complexity of the theory of adaptation which makes it so difficult to comprehend? As even biology majors and medical students struggle to conceptualize these complex theories of evolution (Anderson; Brumby; Mayr; Balgopal; Nehm and Reilly).

Many studies document students' failure to conceptualize evolutionary concepts. For example, Anderson's (2003) study revealed that many nonbiology majors were unable to comprehend the concept of evolution. Her findings exposed student's lack of understanding in the areas of: (1) *random origin of variation*, (2) *how new species originate*, and (3) *how populations change over time*. Anderson's findings are not unique; other researchers document similar findings. In 1984, Brumby examined medical students' reasoning patterns related to natural selection and discovered that they had a poor understanding of the natural selection concept. Students believed that evolutionary changes transpire due to "*need*" within a population (Anderson et al.; Brumby; Stallings). Clough and Wood-Robinson tested young students (ages 12-16) on biological adaptation and discovered that many students had difficulty explaining this concept. The majority of their students used teleological and anthropomorphic explanations (Clough and Robinson; Stallings).

Other studies, like Bishop's and Anderson's ("Student conceptions") evaluated student's knowledge of evolutionary theory after one week of instruction in a 10-week course. The authors reported that students tended to have difficulties in *origin and survival of new traits, the role of variation in a population and evolution as a changing proportion of alleles* (Anderson et al.; Bishop and Anderson, "Student conceptions"; Stallings). Nonbiology students in this study understood the concept of change over time; however, their thinking differed in biological theory or natural selection (Stallings). In 1990, Scharmann used a diversified instructional strategy to evaluate the effectiveness in teaching college students evolutionary concepts. The study showed that diversified strategies were superior to traditional methods (Scharmann; Stallings). Students in Scharmann's study also reported a "feeling of relief" when they discovered that they were not alone in their confusion about the conflicting claims regarding evolutionary concepts (Scharmann; Stallings). Settlage reported a significant decrease in teleological or Lamarckian explanations on student's posttest; the percentage dropped from greater than 50 percent on pretests to less than 20 percent on posttests (Settlage). Therefore, recommending that evolutionary theory be taught year-round instead of limiting it to a single-block of time, unit, or textbook section (Stallings).

Demastes, Settlage and Good closely duplicated Bishop's and Anderson's ((“Student conceptions”) study by using the same conceptual-change teaching module in two separate studies. Study A examined nonbiology majors and Study B investigated high school students. The results of Study A demonstrated that neither the amount of prior instruction nor students' beliefs of evolution were significant factors in the students' use of scientific conceptions (Demastes et al.; Stallings). However, in Study B, the experimental group documented significant use of scientific conceptions (Demastes et al; Stallings). In another study, Demastes, Good, and Pebbles, in an effort to define theory limits, investigated patterns of students' conceptual restructuring of biologic evolution based on conceptual change theory. These researchers illustrated that conceptual change in one sphere requires change in many others and thus, reported conceptual change in: (a) *cascade*, (b) *wholesale*, (c) *incremental*, and (d) *dual constructions* (Demastes et al). Jensen and Finley (“Changes”) also assessed students' learning of evolution by natural selection in four sections of an introductory biology course using different combinations of educational materials (traditional or historically rich materials) and instruction (paired problem solving or traditional lecture). These researchers found few reductions in alternative conceptions even though there were differences between the pre and posttest (Jensen and Finley, “Changes”).

In another study, Reiser, Tabak, Sandoval, Smith, Steinmuller, and Leone reported middle school students' ability to use and comprehend the theory of natural selection. They found that young students are capable of constructing an excellent understanding of natural selection when provided ample time and a cognitively rich learning environment (Anderson; Reiser et al.). Yet in another study, Balgopal reported that students had difficulty understanding: (1) *biotic potential*, (2) *population stability*, (3) *change in a population*, (3) *fitness* and (4) *variation within a population.* Balgopal determined that some students fail to recognize or identify variation caused by chance, which results in differential survival and reproduction rates of organisms (Balgopal).

Furthermore, Nehm and Schonfeld report that even after a 14-week evolution course, about 50% of the teacher participants preferred students be taught some creationism in school (Nehm and Schonfeld, "Does increasing"). They also report that "9% of the biology teachers preferred that students believe creationism exclusively, …[while] 48% of the biology teachers preferred that students believe both evolution and creationism" (Nehm and Schonfeld, "Does increasing" 712). Thus, it is not surprising that many students continue to arrive at institutions of higher education confused and with evolutionary theory misconceptions.

Evolution and Natural Selection: Alternative Conceptions

Alternative conceptions about evolutionary theory, in particular about natural selection, continue to persist. Apparently, instruction often fails to identify and address both mechanical processes and the meaning of scientific terminology, which often differs from general or common use definitions (Anderson; Balgopal).

Evolutionary terms that are currently being used to teach evolutionary theory tend to cause confusion for some students. This problem is not unique in U.S. students. Similar problems have been documented worldwide (Balgopal). For example, Bizzo reported Brazilian high school students ready to enter higher education having difficulty recognizing the differences between biological *competition* and *fighting* (Balgopal). Evolutionary terms that have a tendency to confuse students are words like *adaptation, biological evolution, competition, fitness, belief,* etc.. These words have colloquial definitions that differ from scientific definitions (Anderson et al.; Balgopal; Bishop and Anderson, "Student Conceptions"; Demastes et al.; Rowe). A perfect example is the word *fitness*, which is often referred to or defined as physical strength rather than an organism's ability to survive and reproduce (Anderson et al.; Balgopal). Semantics impact how words are interpreted and understood (Anderson et al.; Balgopal). Furthermore, different languages have different semantic features that are above lexical semantics (Swoyer). Hence, students from diverse populations, for example, Latino students, often bring with them a different language to the learning experience possibly causing them to comprise, interpret, or conceptualize scientific words differently. Consequently, students may arrive at institutions of higher education with pre-conceived notions or misconceptions based on different learning experiences or language usage (Anderson; Balgapol; Demastes et at.). Students may also have colloquial term definitions or explanations that are at odds with scientific meaning (Anderson; Balgapol; Demastes et at.). As a result, students may fail to conceptualize, comprehend, and understand the various processes of natural selection i.e., mutations, variation, adaptation, etc., which are the driving forces behind evolution (Anderson; Balgapol; Demastes et at.).

The scientific research community has been studying metacognition, teaching tools, and learning strategies that encourage conceptual change (Balgopal). Tao and Gunstone suggest that studying the process to understand conceptual change is essential (Tao and Gunstone). After all, "Being able to identify the triggers of conceptual change would allow teachers to integrate intervention strategies into…curricula, promoting metacognition and scientifically accepted concept replacement" (Balgopal 41).

Conceptual Change Theory

Concept learning requires metacognitive and self-reflective capabilities (Balgopal, 2007). Conceptual change refers to an individual's ability to observe and evaluate relationships between concepts and formulate knowledge that can be used to explain situations, address problems, or make educated predictions

(Anderson; Ausubel; Novak and Growin). Conceptual change modifies or transforms an existing conception (e.g., belief, idea, or way of thinking) (Orey). Conceptual change also requires an understanding of learning theories, specifically how knowledge is acquired, organized, and conceptualized. Teaching for conceptual change calls for the uncovering of student's preconceptions about either a topic or phenomenon and requires the use of various teaching techniques/methods to assist the leaner in changing his/her conceptual framework (Davis). It is believed that conceptual change takes place when a learner shifts his/her understanding of reasoning and restructures existing knowledge, beliefs, or conceptions into new foundations of knowledge to solve problems or explain situations (Orey). According to Balgopal, "A student's current understanding and all existing knowledge is referred to as his/her conceptual ecology and encompasses (a) prior knowledge, (b) relationships between concepts, (c) new knowledge about alternative conceptions, and (d) epistemological beliefs" (Balgopal 42). However, according to the literature, not all students are able to undergo conceptual change because some students are not prepared to or know how to either resolve cognitive conflict or learn the meaningful strategies and processes needed to resolve conceptual dilemmas (Balgopal; Hewson and Hewson; Martin, Mintzes, and Clavijo; Mason; Mintzes, Wandersee, and Novak; Novak and Gowin).

Extrapolating from Anderson's work, it is important to reemphasize the role that constructivism and Schema Theory play in conceptual change. The philosophy behind constructivism is based on the notion that students actively perceive and process information to formulate complicated knowledge structures based on what the student already knows and understands (Anderson). Therefore, individual learners must *actively* build knowledge and skills by either adjusting or modifying mental frameworks/representations to accommodate, adapt, or accept to new experiences (Huitt; Orey). Constructivism in itself is multi-faceted and brings to light the importance of identifying and understanding what is possible for students to learn (Anderson; Vygotsky). Vygotsky's original zone of proximal development (ZPD) exemplifies differences between what a leaner is able to do "with" assistance as compared to "without" assistance (Anderson). This pedagogical framework reinforces the concept that students can only build new knowledge or mental frameworks by adding or modifying existing ones (Anderson).

Schema Theory provides the theoretical framework for understanding the cognitive process and proposes the notion that a learner first structure new obtained knowledge into understandable formats by dismantling the information into manageable categories (Rumelhart and Ortony). However, the most important concept of Schema Theory is the role prior knowledge plays during the processing stages of learning (Windmayer). Furthermore, schema plays a significant function in how information is interpreted and decoded, for example, schemata can be reflected in text structures (Driscoll, Halliday and Hassan). Thus, the format in which learners learn is similar to Piaget's model; where a learner can react differently to new knowledge by *accretation, tuning*, and *restructuring* the new information (Windmayer). Therefore, how knowledge is presented directly affects how knowledge is interpreted, coded, categorized, organized, and eventually used. In addition, knowledge structuring is neither global nor domain specific; explaining to some degree why apprentice learners tend to hold different knowledge views which are difficult to modify or change (Vosniadou and Brewer). It is evident that Schema Theory has been expanded and modified since the theory was first introduced by Piaget in 1952. To better understand cognitive learning different measuring methods have had to be developed (Anderson; Stalling).

Levels of Reasoning

The cognitive learning literature strongly suggests that a student's ability to understand complex concepts is directly attributed and correlated to his or her level of reasoning and thus, measures of formal operational reasoning are *highly related to a student's academic achievement* in biology, mathematics, English, and social studies (Lawson; Stallings). Scientific reasoning is defined as:

> ...the ability to logically solve problems through the application of the science method which includes: Problem identification/observation; inductive and deductive reasoning; hypothesis generation; experimentation; interpretation of results; making logical conclusions and critical evaluations...[by] mak[ing] observations and identif[ing] a problem, classif[ing] and interpret[ing] data, develop[ing] a hypothesis, design[ing] experiments/collect data, [and] critically evaluat[ing] experimental outcomes. (Limbaugh 3)

Hence, many studies have looked at how students' reasoning levels correlate to scientific reasoning, biological misconceptions, and scientific beliefs (Lawson and Thompson, 1988). For example, a study of seventh grade students looked at formal reasoning levels on genetics, natural selection, and at student misconceptions. The study's results demonstrated that the only variable strongly associated with students' misconceptions was his/her level of formal reasoning (Lawson and Thompson).

Lawson and Weser reported that nonbiology majors who were less skilled reasoners were not only less likely to change their perceived conceptions/ideas but, were also less likely to commit to scientific beliefs or forms of reasoning (Lawson and Weser; Stallings). Later in 1992, Lawson and Worsnop reported reflective reasoning to be positively correlated to scientific beliefs; however, no correlation was found between reflective reasoning skills and changes in religious beliefs even after evolutionary theory instruction (Lawson and Worsnop).

Learning and Teaching

It only takes a glimpse of the literature to quickly discover and conclude that: (1) as learners construct conceptual understanding or knowledge about a subject, it may differ slightly from one student to the other; (2) no student is a blank slate and thus individual students possess a diverse set of alternative conceptions or notions about the natural world which ultimately influence his/her future learning capabilities; (3) all students, regardless of their ethnicity, gender, age, culture, etc., are capable of conceptualizing, understanding , comprehending, and conducting science; (4) student's construction of knowledge is a process that incorporates different stages: creation, modification, improvement, restructuring; and (5) rejections and newly constructed knowledge are based on the student's diverse sets of experiences and occurrences (Anderson). Many educators believe that the best way for students to learn a concept or idea is by having the learner construct his or her own knowledge structure rather than having someone else construct it for them (Nondestructive Testing, n.d.).

Conceptual change in the sciences has been a pedagogical goal among science educators since the 1970s and even though strong arguments have been made about traditional teaching methods and techniques (e.g., reading, observation of scientific principles, or limited hands-on activities), little has been done to change or diversify traditional instructional methods (Watson and Konicek). The pedagogical literature indicates that as early as the 1920s, the philosophy of science was recommended to be taught as an investigation (Watson and Konicek). However the texts and curricula of the 1950s tell a different story (Watson and Konicek). Even today, many science texts are nothing more than reading books containing predigested demonstrations of various facts asking students questions like, "Does air have weight"; thereby, encouraging students to memorize responses as opposed to assisting them to conceptualize concepts and ideas (Watson and Konicek).

It is not surprising, therefore, that many methods have been employed to measure conceptual knowledge or conceptual change. Scholars have been and are still seeking to assess quantitative and qualitative research techniques to better evaluate student performance, conceptual knowledge, and understanding of evolutionary theory (Balgopal). Instruments range from written analysis discourse, to conceptual mapping, diagnostic short answer tests, diagnostic multiple-choice surveys, student interviews, classroom observations, and laboratory reports and summaries (Balgopal). Even though written reports or reflections provide a glimpse into conceptual knowledge and conceptual change, this method is daunting and time-consuming. Similarly, student interviews are time-consuming, difficult to schedule, and sometimes intimidating to students (Anderson; Balgopal).

Socio-cultural Theory

Socio-cultural theory not only draws on Vygotsky's work but also on other theoreticians such as Tharp and Gallimore who provide a socio-cultural perspective with profound implications in teaching, schooling, and education (Valenzuela). Vygotsky claimed that mental functioning is derived from social interactions and therefore, in order to understand an individual, his/her social context needs to be studied (Vygotsky).

There are two particularly noteworthy aspects to a Vygotskian approach to social interactions. First, it is fundamentally cultural... cultural knowledge...[is an] expression... meaningful within the human sphere of their own culture...Clearly, socio-cultural theory is much more complex than this brief description might lead one to believe. Nonetheless, the aspects described above are important components to consider when examining the communicative and cognitive development of learners...(Valenzuela).

A key idea of socio-cultural theory is constructed by the individual through his/her interactions with others (Balgopal; Mason). Dagher and Crossman also suggested that students' conceptual knowledge is strongly influenced by others so "when teachers and the media use teleological and anthropomorphic language when discussing biological evolution, students learn to do the same" (Balgopal 22-23). Moreover, social interactions direct a "step-by-step" process related to cognitive development, conceptual knowledge/changes, and behavior due to the diffusion of cultural factors (e.g. attitudes, beliefs, actions, activities, conduct, governing factors, etc.) that are either acquired or adapted and vary extensively from culture to culture as cultural practices and cognition constitute one another (Nisbett and Norenzayan).

Methods

Instrument and Procedures:

The paper-and-pencil scantron survey utilized in this research was a modified version of the CINS, which was developed, field-tested, and validated by Anderson, Fisher, and Norman (2002). Permission to modify the CINS was granted by Dr. Dianne L Anderson. In order to keep the questionnaire short, concise and succinct, the total number of pages was limited to five. A shorter format avoids the pitfalls identified by Borg and Gall (1983) who found that "… on average, each page added to the total questionnaire reduced the number of responses by about .5 percent" (Borg and Gall 422).

The first section of the survey contained 10 multiple–choice questions that examined students' conceptual understanding of natural selection. One scenario (Galapagos Finches) was used to assess the theory of natural selection; therefore, question nine and ten were modified by replacing the word "guppies" with the word "finches." By doing this, students were tested only once on the 10 key concepts of natural selection. The key concepts are: (1) *biotic potential*, (2) *population stability*, (3) *natural resources*, (4) *limited survival*, (5) *variation within a population*, (6) *variation inheritable*, (7) *differential survival*, (8) *change in population*, (9) *origin of species*, and (10) *origin of variation* (Anderson et al.).

The second section of the survey contained a series of demographic, sociopolitical, and socio-cultural questions. The demographic variables included: (1) *ethnicity*, (2) *gender*, (3) *age*, (4) *religious preference*, (5) *religiosity*, (6) *student's work status*, (7) *student's income and parents' combined income*, (8) *hometown location*, and (9) *father's and mother's education level*. (1) *Ethnicity*: ethnicity was measured as Latino, Non-Latino (used Latino for those born in U.S. only sample); (2) *Gender*: gender was measured male, female; (3) *Age*: age was measured as a continuous variable; (4) *Religious preference*: religious preference was measured as Catholic, Protestant, non-Christian, other; (5) *Religiosity*: religiosity was measured as never, <once a year, 1-2 year, several times/year, once a month, 2-3 times a month, nearly every week, every week, several times a week, service attendance other than weddings or funerals; (6) *Student's work status*: work status was measured as do work or do not work; (7) *Student's income* and *parents' combined income*: student's and parent's combined income was measured as <$9,999, $10,000-24,999,$25,000-49,999, $50,000-$74,999, $75,000-109,999,>110,000; 8) *Hometown location*: hometown location was measured as an open variable (hometown location indicates whether the student grew up in a metropolitan or nonmetropolitan area. Metropolitan areas are defined as an area with a population of a million or greater (U.S. Census Bureau); and (9) *Father's* and *Mother's education level*: Parent's educational level was measured as <high school, high school (with a diploma or equivalent), <college (with or without a high school diploma or equivalent), technical school completion and/or college attendance, college (college degree), graduate (graduate or professional).

Sociopolitical variables included: (1) *environmental association*, (2) *political affiliation*, (3) *voting practices*, and (4) *political position regarding environmental issues*. (1) *Environmental association*: environmental association was measured as active, sympathetic, neutral, unsympathetic, don't know; (2) *Political affiliation*: political affiliation was measured Republican, Democrat, Independent, other; (3) *Voting practices*: voting practices were measured as voted in the last election (yes, no); and (4) *Political position regarding environmental issues*: political position regarding environmental issues was measured as very important, somewhat important, and not very important.

Researchers use different socio-cultural variables to assess population cultural orientation. In this research project, the variables that were used focused on *ethnicity, ethnic orientation, and acculturation*: (1) *Generation* and (2) *Acculturation* were the two socio-cultural variables used. (1) *Generation*: generation was measured by assessing the number of grandparents born in the U.S.; (2) *Acculturation*: acculturation was evaluated using the Acculturation Rating Scale for Mexican Americans-II (ARMSM-II) which assesses multidimensional acculturative types and measures cultural orientation toward Mexican-American and Anglo-American culture independently (Cuéllar et al.; Lopez). Acculturation was measured using acculturation levels (Level I-very Mexican oriented; Level II, Mexican oriented to approximately balanced bicultural; Level III, slightly Anglo oriented, bicultural; Level IV, strong Anglo oriented; and Level V, very assimilated, Anglicized) (Cuéllar, Arnold, and Maldonado; Cuéllar, et al.).

The modified version of the CINS questionnaire was administered to 372 students enrolled at four Hispanic-Serving Institutions (HIS) in south and west Texas during the 2007 fall semester. The surveys were administered during normal course hours the week before finals. The number of students per class ranged from 24 to 77. Students were given a project information sheet in addition to being informed about the project the day the surveys were administered. Furthermore, students were asked to read the project information sheet before completing the survey. Also, student were informed that participation was on a voluntary basis and anonymous. To ensure anonymity, they were asked not to write their names anywhere on the surveys. Students were not monetarily compensated nor did they earn any bonus points towards their course grade for participation.

Conceptual understanding was evaluated comparing correct response rates against incorrect responses of the two groups (Mexican American and Non-Hispanic Whites). Fifty-two percent of the students who participated were Mexican-Americans compared to forty-six percent who were Anglo-Americans. Sixty-six percent of the participants were females and thirty-three percent were males.

The Sample

Universities:

The following criterion was used to identify potential participating universities:

1. the institution had to be a Texas public 4-year university;
2. it had to be a HSI;
3. it had to be a member of the Hispanic Association of Colleges and niversities (HACU);
4. it had to have a biology, science, or ecology department that offered undergraduate four-year science degrees; and
5. it was required to offer an introductory or first semester biology or ecology course on campus the semester that the data was to be collected.

Even though two-year HSIs play a critical role in higher education; they were not included in this study in that many do not teach introductory biology or ecology courses. HSIs were selected because at least 25 percent of total undergraduate full-time equivalents are Hispanic (Laden). It is important to note that these accredited, degree-granting public or private non-profit institutions were not established to serve a particular ethnic student population but are classified strictly by current student enrollment ratios (Santiago). HSIs were selected for this study because the project's target population was Hispanic/Latino students.

Five universities participated in this study: (1) Sul Ross State University, (2) Texas A&M University-Corpus Christi, (3) Texas A&M University-Kingsville, (4) The University of Texas of the Permian Basin, and (5) The University of Texas-Pan American. The two A&M universities are located in the Texas' Gulf Coast region and offer undergraduate and graduate degrees. A&M-Kingsville ranks among the country's top 10 producers of Hispanic engineers and has the only accredited program in natural gas engineering in the United States. A&M-Corpus Christi ranks in the Master's Universities (West) group and has over 8,000 students. The University of Texas-Pan American is located in the Lower Rio Grande Valley in Edinburg, Texas. This university offers undergraduate and graduate degrees and has a student population of over 17,000. Sul Ross State University and The University of the Permian Basin are located in the Big Bend region of the state. Sul Ross is located in Alpine, Texas and offers 48 undergraduate and 27 graduate degrees. The University of Texas of the Permian Basin services student from approximately 161 Texas counties and offers undergraduate and graduate degrees

(Sources: http://en.wikipedia.org/wiki/Sul_Ross_University; http:// www.sulross.edu/pages/3030. asp; http://en.wikipedia.org/wiki/ Texas_A% 26M_University_at_Corpus_Christi; http://www. tamucc.edu/about/facts .html;http://en.wikipedia.org/wiki/Texas_A%26M_University_at_Kingsvillle; http://www.oir.tamuk.edu/Facts/Old/Default.asp; http://en.wikipedia. org/wiki/University_of Texas_-Pan_American; http://ur.utpa.edu/cr/cr-2404/facts.html; http://en.wikipedia.org/.wiki/ University_of_Texas_ of_the_Permian_Basin ; http://www.utpb.edu/campus-information/, n.d.).

Participants:

Hispanic/Latino students enrolled in four-year public HSIs were the primary target population for this research project. Furthermore, participants selected needed to be enrolled in a first-semester introductory biology or ecology course since the College Board and the Advanced Placement Program's (AP) guiding principles for introductory biology courses require that 25 percent of the course time be spend on evolutionary theory (2007 CollegeBoard , 2008-2009). As indicated earlier, the key elements of evolutionary theory are indispensable in introductory biology as they contribute to the framework students need to conceptualize ideas/concepts and obtain knowledge and skills necessary to assimilate course materials into a conceptual and expandable body of knowledge (2007 CollegeBoard, 2008-2009). Hence in order to facilitate conceptual change, biology educators need to first identify students' pre-existing misconceptions, evaluate them, and then strategize and develop a plan to implement a diverse set of instructional techniques that will result in students' conceptual change. Furthermore, teaching abstract concepts in a relevant context can improve students' attitudes towards academic work (Kirshner and Whitson, 1997).

Findings/Discussion

The National Research Council (NRC, *How*, *Transforming*; Nehm and Reilley) and scholars (e.g. Anderson; Balgopal; Nehm and Reilly) have repeatedly documented that lecture-style instruction has limited impact on student learning and on conceptual change (NRC, *How*, *Transforming*; Anderson; Anderson, et al.; Balgopal, Nehm and Riley). Our study reveals similar finding. Despite having completed a biology course students still retained significant misconceptions about evolutionary theory. Furthermore, no significant differences were found between the Latino and Non-Latino students regarding the most common misconceptions. Both groups identified the same alternative conception on each of the 10 key areas of natural selection. Similarly when the two groups were compared by gender the same common misconception was identified. However, observable differences occurred when students were compared by religious affiliation. As illustrated in Table 1, the different religious groups sometimes identified the most common misconception differently than the other groups. For instance the Non-Christian group identified l*earned behaviors are inherited* as the most common misconception in the scientific concept of change in a population while the other groups identified *mutations occur to meet the needs of the population* as the most common misconception.

Table 1

Description of Alternative Conceptions by Religious Affiliation

Scientific Topic	Alternative Conception	Number of Students Selecting Response Alternative					
		Catholic (n=192)	Protestant (n=122)	Non-Christian (n=15)	Other (n=15)	No-Preference(n=4)	No Religion (n=4)
Biotic potential	Populations level off.	32	33	3	3		
	Organisms only replace themselves.	8	4		1		
Natural resources	Organisms can always obtain what they need to survive.	68	31	6	2		
Population stability	Populations decease.	150	103	11	13	1	3
	All populations grow in size over time.	18	12	2	1		
	Populations always fluctuate	16	4	2	1		

	widely/randomly.						
Change in a population	Mutations occur to meet the needs of the population.	66	48	2	4	1	
	Learned behaviors are inherited.	42	27	6	3		
	Changes in a population occur through a gradual change in all members of a population.	46	24	4	2		1
Limited survival	Organisms work together (cooperate) and don't compete.	79	34	2			
	There is often physical fighting among one species (or among different species) and the strongest one wins.	16	15	1			
Origin of variation	Mutations are intentional: an organism tries, needs, or wants to change genetically.	119	65	8	6		2
	Mutations are adaptive responses to specific environmental agents.	24	15	4		1	

Variation inheritable	Traits that are positively influence by the environment will be inherited by offspring.	53	23	5	4		
	When a trait (organ) is no longer beneficial for survival, the offspring will not inherit the trait.	32	23	6	1		
	Traits acquired during an organism's lifetime will be inherited by offspring.	20	7	2			
Origin of species	Organisms can intentionally become new species over time (an organism tries, wants, or needs to become a new species).	115	65	9	5		
	Speciation is a hypothetical idea.	20	8	1	1		
Variation within a population	Variations only affect outward appearances, don't influence survival.	96	60	6	7	1	2
	All members of the populations are nearly	34	10	2	2		

	identical.						
Differential survival	Fitness is equated with strength, speed, intelligence or longevity.	100	58	8	3		1
	Organisms with many mates are biologically fit.	15	7	1			

It is worth mentioning that a large number of students (47%) across the different religious affiliations interpret "*fitness*" as *equating to strength, speed, intelligence, or longevity.* Hence, our findings are supported by the work of other scholars, who have documented that overall students rely on colloquial vocabulary definitions/explanations regarding the nature of evolution and natural selection conflicting with scientific meaning (Anderson; Balgapol; Demastes et al.).

On the other hand, when comparing percent of correct responses by religious affiliation as illustrated by Table 2, the Catholic group consistently had higher percentage scores on all of the 10 key areas of natural selection than any of the other groups. However, when the percent of correct responses was analyzed by ethnicity and gender more significant differences were noted. Table 3 illustrates student responses by ethnicity and gender. With the exception of one question (measuring biotic potential) the Non-Latino group consistently scored higher than the Latino students. Without knowing what transpired during the student's science education (e.g., the type of curriculum taught, pedagogy teaching styles, number of science courses, etc.), it is difficult to accurately determine why scores vary between the Non-Latino counterparts. However, it is more appropriate to attempt to explain possible differences in terms of relevant research findings. For example, the research indicates that many students arrive at educational institutions with their own belief system and communities (e.g., cultural models, cultural understandings, and educational strategies) which in turn impacts their educational learning experience (Ogbu). In addition, these systems of beliefs may vary significantly from student to student. Secondly,"language use may play a significant role in the mediation of cultural differences in a variety of social cognitive phenomena" (Holtgraves and Kashima 73). Thirdly,

> ...language in its abstract sense is a socially shared tool allowing its users to create and exchange meaning. In so doing, the users of any particular language are connected...[however] it does not guarantee complete and mutual understanding. This is because meaning does not reside solely in the message. Rather, language is used to create meaning, and its users must be active participants in this processes (Holtgraves and Kashima74). [Furthermore] ...language is not merely a selection of words to reflect ideas, but a broader sociocultural process in which language is used to exchange textual, political, and interpersonal messages. (Brown, "It Isn't No" 99)

And as stated earlier, different languages have different semantic features that are above lexical semantics (Swoyer). Finally, perhaps the percent differences exist between the Latino and Non-Latino groups due to students' identity which has been determined to affect science learning (Brown "It Isn't No"). Brown ("It Isn't No") defines *identify* as "who a person is interpreted to be in a given context...[which] reflects a broad set of symbols that are constructed in moment-to-moment interactions, over local time, and over a broader sociohistorical context" (Brown, "It Isn't No" 98). Again studies suggest that many of the minority students' under achievements are due to linguistic differences (Brown, "It Isn't No"; Lee and Fradd) along with inappropriateness of the curriculum sometimes used (Brown, "It Isn't No").

Looking at gender differences we found that females scored 24.4-42.8 higher than the males. It appears that the Latino group had more difficulty responding to questions regarding *change in populations* and *origin of variations* as the percent difference was 22.2 and 24.4 respectively. Males were also challenged by the *origin of variation* scientific concept; the male scores were 42.8% lower than the female scores. However, *change in population* was their least challenging scientific concept. Nonetheless, male respondents did experience difficulty responding to the *origin of species* question; barely over thirty percent of the male participants answered this scientific question correctly.

Table 2

Percent of Correct Responses

Scientific Topic	Catholic (n=192	Protestant (n=122)	Non-Christian (n=15)	Other (n=19)
Biotic potential	58.3	30.8	4.7	6.2
Natural resources	51.3	37.9	3.8	7.1
Population stability	53.4	36.7	3.9	6.1
Change in population	52.1	32.4	4.2	11.3
Limited survival	49.0	37.2	6.1	7.6
Origin of variation	46.6	40.8	2.9	9.7
Variation inheritable	50.0	40.0	1.8	8.3
Origin of species	45.5.	39.8	4.1	10.5
Variation w/i a population	47.6	41.3	5.6	5.6
Differential survival	49.4	37.0	3.9	9.7
Mean	50.3	37.4	4.1	8.2
SD	3.5	3.3	1.2	1.9

Table 3

Percent of Correct Responses

Scientific Topic	Ethnicity Non-Latino/a	Latino/a	Diff	Gender Male	Female	Diff	Total
Biotic potential	45.8	54.2	8.4	33.5	66.5	33.0	61.3
Natural resources	50.2	49.8	.0.4	31.1	68.9	37.8	68.4
Population stability	50.2	49.8	0.4	32.2	67.8	35.6	80.4
Change in population	61.1	38.9	22.2	37.8	62.2	24.4	19.7
Limited survival	55.8	44.2	11.6	32.2	67.8	35.6	56.8
Origin of variatio	62.2	37.8	24.4	28.6	71.4	42.8	30.3
Variation inheritable	52.8	47.2	5.6	31.7	68.3	36.6	49.5
Origin of species	55.9.	44.1	11.8	30.2	69.8	39.6	34.7
Variation w/i a population	53.8	46.2	7.6	32.6	67.4	34.8	36.2
Differential survival	50.0	50.0	0.0	30.3	69.7	39.4	44.3
Mean	53.8	46.2	9.2	32.0	68.0	36.0	48.2
SD	4.9	4.9	8.2	2.3	2.3	4.7	17.8

Change in population refers to the unequal abilities of individuals to survive and reproduce; thus, causing gradual changes in a population in which proportion or ratios of individuals with favorable genetic characteristics will increase overtime and through the generations. While *origin of variation* refers to random mutations and sexual reproductions that transpire and produce genetic variations bestowing survival advantages in some individuals that prove to be beneficial in certain environments, *origin of species* refers to the scientific notion that isolated populations may change drastically overtime and become a new species.

Our results do not differ from Anderson who reported student's lack of understanding in: (1) *random origin of variation*, (2) *how new species originate*, and (3) *how populations change overtime*. Even though we did not conduct a Point Biserial Value and Percent Correct Response analysis, the outcome results are similar to Anderson's findings. Our study reveals that students continue to have difficulty conceptualizing: (1) random *origins of variation, (2) how new species originate, and (3) how populations change overtime*. In addition, many students believe that evolutionary changes transpire due to "*need*" within a population. Both the Latino and Non-Latino groups identified the same misconception regarding *origin of variation,* over 45 percent of the students who answered this scientific question incorrectly believe that *mutations are intentional as organisms want or have the need to change genetically* (Anderson et al., 2002; Brumby; Stallings). Females also identified a similar misconception that *organisms are always able to obtain what they need to survive.* Therefore, some of our findings are not unique to Anderson's or Brumby's studies as Nehm and Schonfeld, "Does Increasing") reported common misconceptions held by science teachers: (1) *mutations caused primarily by mutagenic substances;* (2) *needs cause evolutionary changes to take place*; (3) *the use or disuse of traits explains their appearance/disappearance*; (4) *traits appear only when they are needed*; (5) *all individuals in a population develop new traits simultaneously* (Nehm and Schonfeld, "Does Increasing").

Conclusion and Recommendations

In conclusion, religion as part of an individual's culture serves as the core of an individual's belief system, religion is "*not value free,*" it impacts how students view and come to understand evolution through the natural selection process. In addition, "if science educators are properly concerned with dismantling erroneous conceptions of the nature of science, they may also need to recognize how the prestige of their subject may be precariously poised on the basis of such erroneous conceptions" (Francis and Greer). Despite fears and concerns about lessening individuals' faith through the teachings of evolutionary theory, Francis and Greer study revealed that Catholic and Protestants teenagers living in Northern Ireland could combine positive attitudes towards science and Christianity (Francis and Greer).

Furthermore, as stated earlier, colloquial interpretation of scientific words denote unacceptable scientific use of evolutionary language terms causing students unique challenges. Hence, incorrect definitions of fundamental evolutionary vocabulary impede conceptual knowledge and/or conceptual change. Therefore, a new language should not be used unless it is tied to the consciousness of the old language (Brown, "It Isn't No"). Pedagogy instruction needs to change to include more than a vocabulary list of terms and a few hands-on activities as it needs to "examine students' perceptions of their experience in science classrooms by exploring of their perceptions of the cultural practices of science, the epistemology of science , and the role of discourse in science education" (Brown, "It Isn't No"106). Because, even when instructors/teachers are aware and claim to be sensitive to a diverse student population, they view science as *value neutral* and unrelated to *ethnicity differences* and thus instruction remains traditional and focuses on transmission of scientific knowledge (Brand and Glasson). So it seems feudal to document students' science language discourse when asked about their science experiences. As it does not seem to matter when students clearly state that biology is challenging due to the uses of the same language vocabulary to mean different scientific meanings. Or like one student stated,"...science discourse is of a completely different genre compared to those used in everyday life" (Brown, "It Isn't No").

Indeed, the goal to create conceptual change in the area of natural selection continues to challenge researchers and educators; as natural selection remains as one of the most misunderstood concepts of evolutionary theory (McComas). Again, is it because *origin of variation* requires a sound understanding of genetics (Balgopal) or is it because traditional pedagogy instruction in academia, public, and private secondary schools fails students altogether? The theoretical framework used in this study supports the notion that scientific conceptual change evolves when a learner transitions from one paradigm to another in a nonlinearly fashion (Vosniadou). In addition, the learner replaces existing knowledge and builds skills by adjusting or modifying mental frameworks or mental representations (Huitt; Orey). Hence, in order for educators to facilitate conceptual change; existing pedagogy instruction needs to change so that students do not continue to struggle and fail to "recognize that the origin and the persistence of new traits are controlled by separate mechanisms....[nor be] unable to explain how selective pressures act on variation within a population" (Balgopal 8).

Even though, we did not find significant ethnic differences, more detailed language and culture analysis need to be carried out. Additionally, the identified religious differences prompt interesting questions that demand further investigation. Similarly, are students being handicapped by the different pedagogy instructional styles? Is a student's ability to conceptualize complex theories impaired by the biology teachers' beliefs? The scope of this study does not allow us to answer these questions.

Works Cited

Abd-El-Khalick, F., and Lederman, N.G. "The Influence of History of Science Courses on Students' Views of Nature of Science." *Journal of Research in Science Teaching* 37.10 (2000): 1057-95.

Alao, S., and Gunthrie, J.T. "Predicting Conceptual Understanding with Cognitive and Motivational Variables." *The Journal of Educational Research* 92.4 (1999): 242-54.

Alters, B.J., and Nelson, C.E. "Perpective: Teaching Evolution in Higher Education." *International Journal of Organic Evolution* 56.10 (2002): 1891-1901.

Anderson, D.L. *Natural Selection Theory in Non-Majors' Bbiology: Instruction, Assessment, and Conceptual Difficulty.* ProQuest Digital Dissertations. 2003. AAT 3112820.

Anderson, D.L., Fisher, K.L., and Norman, G.J. "Development and Evaluation of the Conceptual Inventory of Natural Selection." *Journal of Research in Science Teaching* 39.10 (2002): 952-78.

Ausubel, D.P. *Educational Psychology: A Cognitive View.* New York: Holt, Rinehart, and Winston, 1968.

Balgopal, M.M. *Examining Undergraduate Understanding of Natural Selection and Evolution.* ProQuest Digital Dissertations. 2007. AAT 3266974.

Bishop, B.A., and Anderson, C.W. *Occasional Paper No. 91, Evolution by Natural Selection: A Teaching Module.* East Lansing: The Institute for Research on Teaching, Michigan State University, 1986.

---. "Student Conceptions of Natural Selection and Its Role in Evolution." *Journal for Research in Science Teaching* 27.5 (1990): 415-27.

Bizzo, N.M.V. "From Down House Landlord to Brazilian High School Students: What Has Happened to Evolutionary Knowledge on the Way?" *Journal of Research in Science Teaching* 31.5 (1994): 537-56.

Blackwell, W.H., Powell, M.J., and Dukes, G.H. "The Problem of Student Acceptance of Evolution." *Journal of Biological Education* 37.2 (2003): 58-67.

Borg, W.R., and Gall, M.D. *Educational Research: An Introduction.* 4th ed. New York: Longman, 1983.

Brand, B.R., and Glasson, G.E. "Crossing Cultural Borders into Science Teaching: Early Life Experiences, Racial and Ethnic Identities, and Beliefs about Diversity." *Journal of Research in Science Teaching* 41.2 (2004): 119-41.

Brown, B.A. "Discursive Identity: Assimilation into the Culture of Science and Its Implications for Minority Students." *Journal of Research in Science Teaching* 41.8 (2004): 810-34.

----. "It Isn't No Slang That Can Be Said about This Stuff": Language, Identity, and Appropriating Science Discourse." *Journal of Research in Science Teaching* 43.1 (2006): 96-126.

Brickhouse, N. "Bringing In the Outsiders: Reshaping the Sciences of the Future." *Journal of Curriculum Studies* 26.4 (1994): 401-16.

Brumby, M.N. "Misconceptions About the Concept of Natural Selection by Medical Biology Students." *Science Education* 68.4 (1984): 493-503.

Clough, E.E., and Wood-Robinson, C. "How Secondary Students Interpret Instances of Biological Adaptation." *Journal of Biological Education*, 19.2 (1985): 304-10.

College Board. *The College Board.* (2008-2009). Retrieved August 11, 2008, from http://www.collegeboard.com/prod_downloads/ap/students/ biology/ ap-cd-bio-0708.pdf

Cuéllar, I., B. Arnold, and R. Maldonado. "The Acculturation Rating Scale for Mexican Americans-II: A Revision of the Original ARSMA Scale." *Hispanic Journal of Behavioral Science* 17.3 (1995): 275-304.

Cuéllar, I., et al. "Ethnic Identity and Acculturation in a Young Adult Mexican-Origin Population." *Journal of Community Psychology* 25.6 (1997): 535-49.

Dagher, Z.R., and S. Boujaoude. "Students' Perceptions of the Nature of Evolutionary Theory." *Science Education* 89.3 (2005): 378-91.

Dagher, Z.R., and G. Crossman. "Verbal Explanations Given by Science Teachers: Their Nature and Implications." *Journal of Research in Science Teaching* 29.4 (1992): 361-74.

Davis, J. "Conceptual Change." Ed. M. Orey. *Emerging Perspectives on Learning, Teaching, and Technology*. 2001. Date Viewed July 31, 2008, from http://www.coe.uga.edu/epltt/conceptualchange.htm.

Demastes, S.S., J.Jr.Settlage, and R.Good. "Students' Conceptions of Natural Selection and Its Role in Evolution: Cases of Replication and Comparison." *Journal of Research in Science Teaching* 32.5 (1995): 535-50.

Demastes, S.S., R.Good, and P. Peebles. "Patterns of Conceptual Change in Evolution." *Journal of Research in Science Teaching* 33.4 (1996): 407-31.

Driscoll, M. *Psychology of Learning for Instruction*. Boston: Allyn and Bacon, 1994.

Francis, L.J., and J.E. Greer. "Shaping Adolescents' Attitudes Towards Science and Religion in Northern Ireland: The Role of Scientism, Creationism, and Denominational Schools." *Research in Science and Technological Education* 19.1 (2001): 39-53.

Gilbert, A. and R. Yerrick. "Same School, Separate Worlds: A Sociocultural Study of Identity, Resistance, and Negotiation in a Rural, Lower Track Science Classroom." *Journal of Research in Science Teaching*. 38.5 (2001): 574-98.

Halliday and Hassan. *Language, Context, and Text: Aspects of Language in a Social-Semiotic Perspective*. Oxford: Oxford UP, 1989.

Hestenes, D., M.Wells, and G. Swackhamer. "Force Concept Inventory." *The Physics Teacher 30* (1992): 141-58.

Hewson, M.G., and P.W. Hewson. "Effects of Instruction Using Students' Prior Knowledge and Conceptual Change Strategies on Science Learning." *Journal of Research in Science Teaching* 20.8 (1983): 731-43.

Holtgraves, T.M. and Kashima, Y. "Language, Meaning, and Social Cognition." *Personality and Social Psychology Review*, 12 .1 (2008): 73-94.

Huitt, W. "Constructivism." *Educational Psychology Interactive*. Valdosta, GA: Valdosta State University. 2003. Retrieved August 16, 2008, from http://chiron.valdosta.edu/whuitt/col/cogsys/construct.html

Jensen, J.R. *Introductory Digital Image Processing: A Remote Sensing Perspectiev*. 3[Rd] ed. Alexandria, VA: Prentice Hall, 2005.

Jensen, M.S., and F.N. Finley. "Teaching Evolution Using Historical Arguments in a Conceptual Change Strategy." *Science Education* 79.2 (1995): 147-66.

---. "Changes in Students' Understanding of Evolution Resulting from Different Curricular and Instructional strategies." *Journal of Research in Science Teaching*, 33.8 (1996): 879-900.

Kirshner, D. and J.A. Whitson. "Editor's Introduction to Situated Cognition: Social, Semiotic, and Psychological Perspectives." *Situated: Social, semiotic, and psychological perspective*. Ed. D. Kirshner and J.A. Whitson. London, England: Lawrence Erlbaum Associates, 1997. 1-16.

Kutschera, U., and Niklas, K.J. "The Modern Theory of Biological Evolution: An Expanded Synthesis." *Naturwissenschaften* 91.6 (2004): 255-276.

Laden, B.V. "Hispanic-Serving Institutions: Myths and Realities." *Peabody Journal of Education* 76.1 (2001): 73-92.

Lee, O. and S. Fradd. "Science for All, Including Students from Non-English-Language Backgrounds." *Educational Researcher* 27.4 (1998): 12-21.

Lawson, A.E. "A Review of Research on Formal Reasoning and Science Teaching." *Journal of Research in Science Teaching* 22.7 (1985): 569-628.

Lawson, A.E., and L.D. Thompson. "Formal Reasoning Ability and Biological Misconceptions Concerning Genetics and Natural Science." *Journal of Research in Science Teaching* 25.9 (1988): 733-46.

Lawson, A.E., and J. Weser. "The Rejection of Nonscientific Beliefs About Life: Effects of Instruction and Reasoning Skills." *Journal of Research in Science Teaching* 27.6 (1990): 589-606.

Lawson, A.E., and W.A.Worsnop. "Learning About Evolution and Rejecting a Belief in Special Creation: Effects of Reflective Reasoning Skills, Prior Knowledge, Prior Beliefs and Religious Commitment." *Journal of Research in Science Teaching* 29.2 (1992): 143-66.

Lederman, N.G. "Students' and Teachers' Conceptions of the Nature of Science: A Review of the Research." *Journal of Research in Science Teaching* 29.4 (1992): 331-59.

Libarkin, J.C., and S.W. Anderson. (2005). "Assessment of Learning in Entry-Level Geoscience Courses: Results from the Geoscience Concept Inventory." *Journal of Geoscience Education* 53.4 (2005): 394-401.

Limbaugh, J. *Basic Proficiencies in Scientific Reasoning*. 2005. Retrieved August 3, 2008, from www.frostburg.edu/academic/slassessment /final%20recommendation%20SCIEN%20REASONING.doc

Lopez, A. *Texas Latino knowledge and attitudes towards natural resources and the environment.* 2005. Retrieved from Texas AandM Thesis and Dissertations. (http://handle.tamu.edu/1969.1/3238).

Martin, B.L., J.J Mintzes, and I.E. Clavijo. "Restructuring Knowledge in Biology: Cognitive Processes and Metacognitive Reflections." *International Journal of Science Education* 22.3 (2000): 303-323.

Mason, L. "Sharing Cognition to Construct Scientific Knowledge in School Context: The Role of Oral and Written Discourse." *Instructional Science* 26.5(1998): 359-89.

Mayr, E. *The Growth of Biological Thought: Diversity, Evolution, and Inheritance*. Cambridge, MA: Harvard UP, 1982.

McComas, W. F., ed. *Investigating Evolutionary Biology in the Laboratory.* Reston, VA: National Association of Biology Teachers, 1994.

Mintzes, J.J., J.H. Wandersee, and J.D. Novak. *Assessing Science Understanding: A Human Constructivist View.* San Diego, CA: Academic Press, 2000.

Montgomery, A., and W.K. Kay. "Relationship Toward Assembly and Religious Education Among Roman Catholic Girls." *Research in Religious Education.* Ed. L.J. Francis, W.K. Kay, and W.S. Campbell. New York: Gracewing Publishing, 1996. 393-405.

Nehm, R.H., and I.S. Schonfeld "Does Increasing Biology Teacher Knowledge of Evolution and the Nature of Science Lead to Greater Preference for the Teaching of Evolution in Schools?" *Journal of Science Teacher Education* 18.5 (2007): 699-723.

---. "Measuring Knowledge of Natural Selection: A Comparison of the CINS, an Open-Response Instrument, and an Oral Interview." 2008. *Journal of Research in Science Teaching, 9999.* Retrieved July 29, 2008, from http://www3.interscience.wiley.com/journal/119031195/abstract?CRETRY=1andSRETRY=0.

Nehm, R.H., and L. Reilly. "Biology Majors' Knowledge and Misconceptions of Natural Selection." *BioScience*, 7.3 (2007): 263-72.

Nisbett, R.E., and A. Norenzayan. "Culture and Cognition." Ed. D. Medin, and H. Pashler. *Memory and Cognitive Processes.* Vol. 2 of *Stevens' Handbook of Experimental Psychology.* 3rd ed. New York: John Wiley and Sons, 2002. 561-97.

National Research Council. *How People Learn: Brain Mind Experience, and School.* Exp.ed. Washington, DC: National Academies Press, 2000.

---. *Transforming Undergraduate Education for Future Research Biologist.* Washington, DC: National Academies Press, 2003.

Nondestructive Testing. *Teaching with the Constructivist Learning Theory.* n.d. Retrieved August 3, 2008, from http://www.ndt-ed.org/Teaching ResourceConstructivist%20_Learning.htm

Novak, J.D. and D.B. Growin. *Learning How to Learn.* Cambridge: Cambridge UP, 1984.

Ogbu, J.U. "Understanding Cultural Diversity and Learning." *Educational Research* 21.8 (1992): 5-14.

Orey, M., ed. *EmergingPperspectives on Learning, Teaching, and Technology.* 2001. Retrieved July 31, 2008, from http://projects.coe.uga.edu/epltt/

Pidwirny, M. "Natural Selection and Evolution." *Fundamentals of Physical Geography.* 2nd ed. 2006. Retrieved August 15, 2008. http://www.physicalgeography.net/fundamentals/9c.html

Raia, F. "Students' Understanding of Complex Dynamic Systems." *Journal of Geoscience Education* 53.3 (2005): 297-308.

Reiser, B., et al. "BGuILE: Strategic and Conceptual Scaffolds for Scientific Inquiry in Biology Classrooms." Cognition and instruction: twenty-Five years of progress. Ed. S. M. Carver, and D. Klahr. Mahwah, NJ: Erlbaum, 2001. 263-305.

Rowe, M.F. "College Students' Understanding of Natural Selection: Moving From a Lamarckian Perspective to a Darwinian Point of View." Paper presented at the annual meeting of the National Association of Research in Science Teaching. Vancouver, British Columbia, 2004.

Rumelhart, D.E., and A. Ortony. "The Representation of Knowledge in Memory." *Schooling and the Acquisition of Knowledge.* Ed. R.C.Anderson, R.J. Spiro, and E. Montague. Hillsdale, NJ: Erlbaum, 1977. 99-135.

Sadler, T.D. "Evolutionary Theory as a Guide to Socioscientific Decision-Making." *Journal of Biological Education* 39.2 (2005): 68-72.

Santiago, D.A. "INVESTING." *Hispanic-Serving Institutions (HSIs): The Basics.* 2006. Retrieved February 11, 2008, from www.edexcelencia.org/pdf/InventingHSIsFINAL.pdf

Scharmann, L.C. "Enhancing the Understanding of the Premises of Evolutionary Theory: The Influence of a Diversified Instructional Strategy." *School Science and Mathematics* 90.2 (1990): 91-100.

Settlage, J.Jr. "Conceptions of Natural Selection: A Snapshot in the Sense-Making Process." *Journal of Research in Science Teaching* 31.5 (1994): 449-57.

Stallings, M.A. *Enhancing andUunderstanding of Evolution: The Influence of a Diversified Instructional Strategy on High School Biology Students.* 1996. Retrieved from ProQuest Digital Dissertations. (AAT 9714357).

Sul Ross State University. "Sul Ross State University." n.d. Retrieved May 1, 2008, from http://en.wikipedia.org/wiki/Sul_Ross_University http://www .sulross.edu/pages/3030.asp

Swoyer, C. *Stanford Encyclopedia of Physlosophy.* 2003. Retrieved July 8, 2008, from http://plato.stanford.edu/entries/relativism/supplement2.html

Tao, P-K., and R.F. Gunstone. "The Process of Conceptual Change in Force and Motion During Computer-Supported Physics Instruction." *Journal of Research in Science Teaching* 36.7 (1990): 859-82.

Texas A&M University-Corpus Christi. "Texas A&M University-Corpus

Christi." n.d. Retrieved May 1, 2008, from http://en.wikipedia.org /wiki/Texas_A%26M_University_at_Corpus_Christi http://www.tamucc. edu/about/facts.html

Texas A&M University-Kingsville. "Texas A&M University-Kingsville." n.d. Retrieved May 1, 2008, from http://en.wikipedia.org/wiki/ Texas_A%26M_University_at_Kingsvillle http://www.oir.tamuk.edu/ Facts/Old/Default.asp

Tharp, R. G. and R. Gallimore. *Rousing Minds to Life: Teaching, Learning, and Schooling in Social Context.* Cambridge: Cambridge UP, 1989.

The University of Texas at Pan-Am. "The University of Texas at Pan-Am." n.d. Retrieved May 1, 2008, from http://en.wikipedia.org/wiki/ University_of_Texas_-_Pan_American http://ur.utpa.edu/cr/cr-2404/ facts. html

U.S. Bureau of the Census. "Metropolitan Areas." n.d. Retrieved April 27, 2008, from http://www.census.gov/population/www/estimates/metro area.html

University of Texas of the Permian Basin. "University of Texas of the Permian Basin." n.d. Retrieved May 1, 2008, from http://en.wikipedia.org/ wiki/University_of_Texas_of_the_Permian_Basin http://www.utpb.edu/ campus-information/

Valenzuela, J.S. *Sociocultural Theory.* n.d. Retrieved August 4, 2008, from http://www.unm.edu/~devalenz/handouts/sociocult.html

Vosniadou, S. "The Cognitive-Situative Divide and the Problem of Conceptual Change." *Educational Psychologist*, 42.1 (2007): 55-66.

Vosniadou, S., and W.F. Brewer. "Theories of Knowledge Restructuring in Development." *Review of Educational Research* 57.1 (1987): 51-67.

Vygotsky. L.S. *Mind in society: The Development of Higher Psychological Processes.* Cambridge, MA: Harvard UP, 1978.

Watson, B. and Konicek, R. "Teaching for Conceptual Change: Confronting Children's Experiences." *Phi Delta Kappan.* 71.9 (1990): 680-85.

Windmayer, S.A. *Schema Theory: An Introduction.* 2007. Retrieved August 12, 2008, from http://www2.yk.psu.edu/~jlg18/506/SchemaTheory.pdf

Yeo, S., and M. Zadnick. "Introductory Thermal Concept Evaluation: Assessing Students' Understanding." *The Physics Teacher* 39.8 (2001): 496-504.

Zeilik, M., C. Schau, and N. Mattern. "Conceptual Astronomy. II. Replicating Conceptual Gains, Probing Attitude Changes across Three Semesters." *American Journal of Physics* 67.10 (1999): 923-27.

SELECTED BIBLIOGRAPHY

Alcorta, Candace S., and Richard Sosis. "Ritual, Emotion and Sacred Symbols: The Evolution of Religion as an Adaptive Complex." *Human Nature* 16 (2005): 323-359.

Alexander, Richard D. *The Biology of Moral Systems*. Hawthorne, NY: Aldine De Gruyter, 1987.

---. "Evolution of the Human Psyche." *The Human Revolution: Behavioural and Biological Perspectives on the Origins of Modern Humans*. Ed. Paul Mellars and Chris Stringer. Princeton: Princeton UP, 1989. 455-513.

Anderson, Joseph. *The Reality of Illusion: An Ecological Approach to Cognitive Film Theory*. Carbondale: Southern Illinois UP, 1996.

---, and Barbara Fisher Anderson, eds. *Moving Image Theory: Ecological Considerations*. Carbondale: Southern Illinois UP, 2005.

Appel, Markus, and Tobias Richter. "Persuasive Effects of Fictional Narratives Increase over Time." *Media Psychology* 10 (2007): 113-34.

Ayala, Francisco J. *Darwin and Intelligent Design*. Minneapolis: Fortress Press, 2006.

---. *Darwin's Gift to Science and Religion*. Washington , D.C : Joseph Henry, 2007.

Barash, David, and Nanelle Barash. *Madame Bovary's Ovaries: A Darwinian Look at Literature*. New York: Delacorte, 2005.

---. "Evolution and Literary Criticism." *The Chronicle of Higher Education* 49:8 (18 October 2002): B7-B9.

Barkow, Jerome, Leda Cosmides, and John Tooby, eds. *The Adapted Mind: Evolutionary Psychology and the Generation of Culture*. New York: Oxford UP, 1992.

Barrett, Louise, Robin Dunbar, and John Lycett. *Human Evolutionary Psychology*. Princeton: Princeton UP, 2002.

Barrish, Phillip. "Accumulating Variation: Darwin's *On the Origin of Species* and Contemporary Literary and Cultural Theory." *Victorian Studies: A Journal of the Humanities, Arts, and Sciences* 34 (1991): 431-53.

Baumeister, Roy F. *The Cultural Animal: Human Nature, Meaning, and Social Life*. Oxford: Oxford UP, 2005.

Bednarik, Robert. "Cupules: The Oldest Surviving Rock Art." *Rock Art Research*, forthcoming.

Bekoff, Marc. *Minding Animals: Awareness, Emotions, and Heart*. New York: Oxford UP, 2002.

---, Colin Allen, and Gordon Burghardt (eds.). *The Cognitive Animal: Empirical and Theoretical Perspectives on Animal Cognition*. Cambridge: MIT P, 2002.

Benzon, William. "Signposts for a Naturalist Criticism." *Entelechy: Mind and Culture* No. 6 (Fall 2005/Spring 2006). http://www.entelechy journal .com/billbenzon.html.

Bickerton, Derek. *Language and Species*. Chicago: U of Chicago P, 1990.

Bjorklund, David F., and Anthony D. Pellegrini. *The Origins of Human Nature: Evolutionary Developmental Psychology*. Washington, D. C.: APA, 2002.

Boehm, Christopher. *Hierarchy in the Forest: The Evolution of Egalitarian Behavior*. Cambridge: Harvard UP, 1999.

Boyd, Brian. "Art and Evolution: Spiegelman in *The Narrative Corpse*." *Philosophy and Literature* 32 (2008): 31-57.

---. "Artistic Animals: Common and Unique Features of Music and Visual Art." *The Visual Animal*. Ed. Ian North. Adelaide: Centre for Contemporary Art, 2007. 15-26.

---. "The Art of Literature and the Science of Literature." *The American Scholar*. 77:2 (2008): 118-27.

---. "Evolutionary Theories of Art." Gottschall and Wilson 147-76.

---. "Getting It All Wrong." *The American Scholar*. 75.4 (2006): 18-30.

---. "Jane, Meet Charles: Literature, Evolution, and Human Nature." *Philosophy and Literature* 22 (1998): 1-30.

---. "Literature and Evolution: A Bio-Cultural Approach." *Philosophy and Literature* 29 (2005): 1-23.

---. *On the Origin of Stories: Evolution, Cognition, and Fiction*. Cambridge: Harvard UP, forthcoming.

---. "The Origin of Stories: *Horton Hears a Who*." *Philosophy and Literature* 25 (2001): 197-214.

---, Joseph Carroll, and Jonathan Gottschall, eds. *Evolutionary Approaches to Literature and Film: A Reader in Art and Science*. Under submission.

Brown, Donald. *Human Universals*. Philadelphia: Temple UP, 1991.

Buller, David. *Adapting Minds: Evolutionary Psychology and the Persistent Quest for Human Nature*. Cambridge, MA: MIT P, 2005.

Burgess, Robert L., and Kevin MacDonald (eds.). *Evolutionary Perspectives on Human Development*. 2nd ed. Thousand Oaks, CA: Sage, 2005.

Burghardt, Gordon M.. "Darwin's Enduring Legacy to Ethology and Comparative Psychology." *American Psychologist*, forthcoming.

Buss, Arnold. "Evolutionary Perspectives on Personality Traits." *Handbook of Personality Psychology*. Ed. Robert Hogan, John Johnson, and Stephen Briggs. San Diego: Academic P, 1997. 346-66.

Buss, David M. *The Dangerous Passion: Why Jealousy Is as Necessary as Love and Sex*. New York: Free Press, 2000.

---. *Evolutionary Psychology: The New Science of the Mind*. 3rd ed. Boston: Allyn and Bacon, 2007.

---. *The Evolution of Desire: Strategies of Human Mating*. 2nd ed. New York: Basic, 2003.

---, ed. *The Handbook of Evolutionary Psychology*. Hoboken, NJ: Wiley, 2005.

Campbell, Bernard, ed. *Sexual Selection and the Descent of Man, 1871-1971.* Chicago: Aldine, 1971.

Carroll, Joseph. "Adaptationist Literary Study: An Emerging Research Program." *Style* 36 (2002): 596-617.

---. "Adaptationist Literary Study: An Introductory Guide." *Ometeca* 10 (2006): 18-31.

---. "The Adaptive Function of Literature." *Evolutionary Approaches to the Arts.* Ed. Paul Locher, et al. Amityville, NY: Baywood, 2007. 31-45.

---. "Aestheticism, Homoeroticism, and Christian Guilt in *The Picture of Dorian Gray*: A Darwinian Critique." *Philosophy and Literature* 29 (2005): 286-304.

---. "The Cuckoo's History: Human Nature in *Wuthering Heights*." *Philosophy and Literature*. Forthcoming.

---. *Evolution and Literary Theory.* U of Missouri P, 1995.

---. "Evolutionary Approaches to Literature and Drama." *Oxford Handbook* Ed. Dunbar and Barrett Oxford: Oxford UP, 2007. 637-48.

---. "Human Nature and Literary Meaning. A Theoretical Model illustrated with a Critique of *Pride and Prejudice*." Gottschall and Wilson 76-106.

---. "The Human Revolution and the Adaptive Function of Literature." *Philosophy and Literature* 30 (2006): 33-49.

---. *Literary Darwinism: Evolution, Human Nature, and Literature.* New York: Routledge, 2004.

---. "Literature and Evolutionary Psychology." D. Buss, *Handbook* 931-52.

---. "Literature as a Human Universal." *Grenzen der Literatur: Zu Begriff und Phänomen des Literarischen.* Ed. Fotis Jannidis, Gerhard Lauer, and Simone Winko. Berlin: de Gruyter, forthcoming.

---. "Steven Pinker's Cheesecake for the Mind." *Philosophy and Literature* 22 (1998): 478-85. Rpt. in *Literary Darwinism: Evolution, Human Nature, and Literature.* New York: Routledge, 2004. 63-68.

---, and Jonathan Gottschall. "Human Nature and Agonistic Structure in Canonical British Novels of the Nineteenth and Early Twentieth Centuries: A Content Analysis." *Anthropologie und Sozialgeschichte der Literatur Heuristiken der Literaturwissenschaft.* Ed. Uta Klein, Katja Mellmann, and Stephanie Metzger. Paderborn, Germany: Mentis, 2006. 473-87.

---, Jonathan Gottschall, John Johnson, and Daniel Kruger. *Graphing Jane Austen: Human Nature in British Novels of the Nineteenth Century.* Under submission.

Clay, Felix. *The Origin of the Sense of Beauty: Some Suggestions upon the Source and Development of the Aesthetic Feelings.* London: Smith, Elder, 1908.

Coe, Kathryn. *The Ancestress Hypothesis: Visual Art as Adaptation.* New Brunswick: Rutgers UP, 2003.

Cooke, Brett. *Human Nature in Utopia: Zamyatin's* We. Evanston: Northwestern UP, 2002.

---. "On the Evolution of Interest: Cases in Serpent Art." *Evolution of the Psyche.* Ed. David H. Rosen and Michael Luebbert. Westport, CT: Praeger, 1999. 150-68.

---. "The Promise of a Biothematics." *Sociobiology and the Arts.* Ed. Jean Baptiste Bedaux and Brett Cooke. Amsterdam: Editions Rodopi, 1999. 43-62.

---. "Sexual Property in Pushkin's 'The Snowstorm': A Darwinist Perspective." Cooke and Turner 175-204.

---, and Frederick Turner, eds. *Biopoetics: Evolutionary Explorations in the Arts.* Lexington, KY: ICUS, 1999.

Cosmides, Leda, and John Tooby. "Consider the Source: The Evolution of Adaptations for Decoupling and Metarepresentation." *Metarepresentations: A Multidisciplinary Perspective.* Ed. Dan Sperber. New York: Oxford UP, 2000. 53-115.

Crews, Frederick. *Out of My System: Psychoanalysis, Ideology, and Critical Method.* New York: Oxford UP, 1975.

---. "Foreword from the Literary Side." Gottschall and Wilson xiii-xv.

---. *Follies of the Wise: Dissenting Essays.* Emeryville, CA: Shoemaker and Hoard, 2006.

---. *Postmodern Pooh.* 2001. Northwestern UP, 2007.

Currie, Gregory. *Arts and Minds.* Oxford: Clarendon, 2004.

---. "The Representational Revolution." *The Journal of Aesthetics and Art Criticism* 62 (2004): 119-28.

Daly, Martin, and Margo Wilson. *Homicide.* New York: Aldine, 1988.

Darwin, Charles. *The Descent of Man, and Selection in Relation to Sex.* 1871. Ed. John Tyler Bonner and Robert M. May. 2 Vols. in 1. Princeton: Princeton UP, 1981.

---. *The Expression of the Emotions in Man and Animals.* London: J. Murray, 1872, p. 166).

---. *On the Origin of Species by Means of Natural Selection.* 1859 Ed. Joseph Carroll. Peterborough, Ontario: Broadview, 2003.

Darwin, Francis, ed. *The Life and Letters of Charles Darwin, Including an Autobiographical Chapter.* Vol. 1. London: John Murray, 1887.

Dawkins, Richard. *The Blind Watchmaker.* New York: W. W. Norton, 1986.

---. *The Extended Phenotype: The Gene as the Unit of Selection.* 1982. New York: Oxford UP, 1983.

---. *The Selfish Gene.* New York: Oxford UP, 1976.

Deacon, Terrence W. *The Symbolic Species: The Co-Evolution of Language and the Brain.* New York: Norton, 1997.

Degler, Carl N. *In Search of Human Nature: The Decline and Revival of Darwinism in American Social Thought.* New York: Oxford UP, 1991.

de Waal, Frans. *The Ape and the Sushi Master: Cultural Reflections by a Primatologist.* New York: Basic Books, 2001.

---. *Primates and Philosophers: How Morality Evolved*. Princeton: Princeton UP, 2006.

Dissanayake, Ellen. *Art and Intimacy: How the Arts Began*. Seattle: U of Washington P, 2000.

---. *Homo Aestheticus: Where Art Comes from and Why*. 1992. Seattle: U of Washington P, 1995.

---. "An Hypothesis of the Evolution of Art from Play." *Leonardo* 7 (1974): 211-18.

---. "In the Beginning: Pleistocene and Infant Psychology and 21st Century Education in the Arts." *International Handbook of Research in Arts Education*. Ed. Liora Bresler. Berlin: Springer, 2006. 781-795.

---. "'Making Special': An Undescribed Human Universal and the Core of a Behavior of Art." *Biopoetics: Evolutionary Explorations in the Arts*. Brett Cooke and Frederick Turner (Eds.). Lexington, Kentucky: ICUS, 1999. 27-46.

---. "The Universality of the Arts in Human Life." *Understanding the Arts and Creative Sector in the United States*. Ed. Joni Maya Cherbo, Margaret J. Wyszomirski, and Ruth Ann Stewart. New Brunswick, NJ: Rutgers UP, forthcoming.

---. "What Art Is and What Art Does: An Overview of Contemporary Evolutionary Hypotheses." *Evolutionary and Neurocognitive Approaches to Aesthetics, Creativity, and the Arts*. Ed. Colin Martindale, Paul Locher, and Vladimir M. Petrov. Amityville, NY: Baywood, 2007. 1-14.

---. *What Is Art For?* Seattle: U of Washington P, 1988.

Donald, Merlin. *A Mind So Rare: The Evolution of Human Consciousness*. New York: Norton, 2001.

Dunbar, Robin. *The Human Story: A Brief History of Mankind's Evolution*. London: Faber, 2004.

---. "Why Are Good Writers So Rare? An Evolutionary Perspective on Literature." *Journal of Evolutionary and Cultural Psychology* 3 (2005): 7-22.

---, and Louise Barrett. "Evolutionary Psychology in the Round." Dunbar and Barrett, *Oxford Handbook* 3-9.

---, and Louise Barrett, eds. *Oxford Handbook of Evolutionary Psychology*. Oxford: Oxford UP, 2007.

Dutton, Denis. *The Art Instinct*. Oxford: Oxford UP, forthcoming.

Easterlin, Nancy. "Do Cognitive Predispositions Predict or Determine Literary Value Judgments? Narrativity, Plot, and Aesthetics." Cooke and Turner 241-62.

---. "Hans Christian Andersen's Fish Out of Water." *Philosophy and Literature* 25 (2001): 251-77.

---. "Loving Ourselves Best of All": Ecocriticism and the Adapted Mind. *Mosaic* 37 (2004): 1-18.

---. “Psychoanalysis and the ‘Discipline of Love.’” *Philosophy and Literature* 24 (2000): 261-79.

Easton, Alexander, and Nathan Emery (eds.). *The Cognitive Neuroscience of Social Behaviour*. New York: Psychology P, 2005.

Eibl-Eibesfeldt, Irenäus. “Us and the Others: The Familial Roots of Ethnonationalism.” *Ethnic conflict and indoctrination: Altruism and identity in evolutionary perspective*. In Ed. Irenäus Eibl-Eibesfeldt and F. K. Salter. New York: Berghahn, 1998. 21-53.

Ellis, Bruce J., and David F. Bjorklund, eds. *Origins of the Social Mind: Evolutionary Psychology and Child Development*. New York: Guilford, 2005.

Emery, Nathan, Nicola Clayton, and Christopher Frith (eds.). *Social Intelligence from Brain to Culture*. New York: Oxford UP, 2008.

Enard, Wolfgang, et al. “Molecular Evolution of *FOXP2*, a Gene Involved in Speech and Language.” *Nature* (2002): 869-72.

Evans, Dylan. “From Lacan to Darwin.” Gottschall and Wilson 38-55.

Fox, Robin. “Male Bonding in the Epics and Romances.” Gottschall and Wilson 126-146.

Fromm, Harold. *From Ecology to Consciousness*. Baltimore: Johns Hopkins UP, forthcoming.

---. “The New Darwinism in the Humanities: From Plato to Pinker.” *Hudson Review* 56 (2003): 89-99.

---. “The New Darwinism in the Humanities, Part Two: Back to Nature Again.” *Hudson Review* 56 (2003): 315-27.

Gangestad, Steven W., and Jeffry A. Simpson. “An Introduction to *The Evolution of Mind*: Why We Developed This Book.” Gangestad and Simpson, *Evolution of Mind* 1-21.

---, and Jeffry A. Simpson, eds. *The Evolution of Mind: Fundamental Questions and Controversies*. New York: Guilford, 2007.

Geary, David C. “Evolution of Paternal Investment.” D. Buss, *Handbook* 483-505.

---. *Male, Female: The Evolution of Human Sex Differences*. Washington, D. C.: APA, 1998.

---. *The Origin of Mind: Evolution of Brain, Cognition, and General Intelligence*. Washington, D. C.: APA, 2005.

---, and Mark V. Flinn. “Evolution of Human Parental Behavior and the Human Family.” *Parenting: Science and Practice* 1 (2001): 5-61.

Goodheart, Eugene. *Darwinian Misadventures in the Humanities*. New Brunswick, NJ: Transaction, 2007.

Gottschall, Jonathan. “An Evolutionary Perspective on Homer’s Invisible Daughters.” *Interdisciplinary Literary Studies* 4 (2003): 36-55.

---. “Homer’s Human Animal: Ritual Combat in the *Iliad*.” *Philosophy and Literature* 25 (2001): 278-94.

---. *Literature, Science, and a New Humanities*. Houndmills: Palgrave MacMillan, 2008.

---. "Measure for Measure." *Boston Globe*. 11 May 11 2008: D1-2.

---. *The Rape of Troy: Evolution, Violence, and the World of Homer*. Cambridge: Cambridge UP, forthcoming.

---. "The Tree of Knowledge and Darwinian Literary Study." *Philosophy and Literature* 27 (2003): 255-68.

---, and Marcus Nordlund. "Romantic Love: A Literary Universal?" *Philosophy and Literature* 30 (2006): 450-70.

---, and David Sloan Wilson, eds. *The Literary Animal: Evolution and the Nature of Narrative*. Evanston: Northwestern UP, 2005.

Hagen, Edward H., and Donald Symons. "Natural Psychology: The Environment of Evolutionary Adaptedness and the Structure of Cognition." Gangestad and Simpson, *Evolution of Mind* 38-44.

Hamilton, William D. "The Genetical Evolution of Social Behavior, I and II." *Narrow Roads of Gene Land: The Collected Papers of W. D. Hamilton*. Vol. 1. Oxford: W. H. Freeman, 1996. 11-82.

Hammerstein, Peter, ed. *Genetic and Cultural Evolution of Cooperation*. Cambridge: MIT P, 2003.

Harris, Judith Rich. *No Two Alike: Human Nature and Human Individuality*. New York: Norton, 2007.

Hassin, Ran R., James S. Uleman, and John A. Bargh, eds. *The New Unconscious*. New York: Oxford UP, 2005.

Hauser, Marc. *Wild Minds: What Animals Really Think*. New York: Henry Holt, 2000.

Headlam Wells, Robin. *Shakespeare's Humanism*. Cambridge: Cambridge UP, 2005.

---, and Johnjoe McFadden, eds. *Human Nature: Fact and Fiction*. London: Continuum, 2006.

Henshilwood, Christopher S., and Curtis W. Marean. "The Origin of Modern Human Behavior: Critique of the Models and Their Test Implications." *Current Anthropology* 44 (2003): 627-51.

Hill, Kim. "Evolutionary Biology, Cognitive Adaptations, and Human Culture." Gangestad and Simpson, *Evolution of Mind* 348-56.

Hobson, Peter. *The Cradle of Thought: Exploring the Origins of Thinking*. Oxford: Oxford University Press, 2004.

Hodgson, Geoffrey M. "Taxonomizing the Relationship between Biology and Economics: A Very Long Engagement." *Journal of Bioeconomics* 9 (2007): 169-85.

Hogan, Patrick Colm. *Cognitive Science, Literature, and the Arts: A Guide for Humanists*. New York: Routledge, 2003.

---. *The Mind and Its Stories: Narrative Universals and Human Emotion*. Cambridge: Cambridge UP, 2003.

Horvath, Tim. "By Sextants and Stars: The Propitious Voyage of *The Literary Animal*. A review of *The Literary Animal: The Evolution and the Nature of Narrative* by Jonathan Gottschall and David Sloan Wilson." *Human Nature Review*. 5 (2005): 99-107.

Hyman, Stanley Edgar. *The Tangled Bank: Darwin, Marx, Frazer, and Freud as Imaginative Writers*. New York: Grosset and Dunlap, 1966.

Jackson, Tony. "Issues and Problems in the Blending of Cognitive Science, Evolutionary Psychology, and Literary Study." *Poetics Today* 23 (2002): 161-79.

---. "Questioning Interdisciplinarity." *Poetics Today* 21 (2000): 319-47.

Jobling, Ian. "Personal Justice and Homicide in Scott's *Ivanhoe*: An Evolutionary Psychological Perspective." *Interdisciplinary Literary Studies* 2 (2001): 29-43.

Kelleter, Frank. "A Tale of Two Natures: Worried Reflections on the Study of Literature and Culture in an Age of Neuroscience and Neo-Darwinism." *Journal of Literary Theory* 1 (2007): 151-87.

Klein, Richard G. *The Human Career: Human Biological and Cultural Origins*. Chicago: U of Chicago P, 1989.

Kirby, Simon. "The Evolution of Language." Dunbar and Barrett, *Oxford Handbook* 669-81.

Kruger, Daniel, Maryanne Fisher, and Ian Jobling. "Proper and Dark Heroes as Dads and Cads: Alternative Mating Strategies in British and Romantic Literature." *Human Nature* 14 (2003): 305-317.

Levitin, Daniel J. *This Is Your Brain on Music: The Science of a Human Obsession*. New York: Dutton, 2006.

Love, Glen A. "Ecocriticism and Science: Toward Consilience?" *New Literary History* 30 (1999): 561-76.

Low, Bobbi S. *Why Sex Matters: A Darwinian Look at Human Behavior*. Princeton: Princeton UP, 2000.

McBrearty, Sally, and Alison S. Brooks. "The Revolution That Wasn't: A New Interpretation of the Origin of Modern Human Behavior." *Journal of Human Evolution* 39 (2000): 453-563.

MacDonald, Kevin. "Evolution, Culture, and the Five-factor Model." *Journal of Cross-Cultural Psychology* 29 (1998): 119-149.

Mayr, Ernst. *Evolution and the Diversity of Life: Selected Essays*. Cambridge: Harvard UP, 1976.

Mellard, James M. "'No ideas but in things': Fiction, Criticism, and the New Darwinism." *Style* 41 (2007): 1-29.

Mithen, Steven. "Mind, Brain, and Material Culture: An Archaeological Perspective." *Evolution and the Human Mind: Modularity, Language, and Meta-Cognition*. Ed. Peter Carruthers and Andrew Chamberlain. Cambridge: Cambridge UP, 2000. 207-17.

---. *The Prehistory of the Mind: The Cognitive Origins of Art, Religion, and Science*. London: Thames and Hudson, 1996.

---. *Thoughtful Foragers: A Study of Prehistoric Decision Making*. Cambridge: Cambridge UP, 1990.

Montell, Conrad. "On Evolution of God-Seeking Mind: An Inquiry into Why Natural Selection Would Favor Imagination and Distortion of Sensory Experience." *Evolution and Cognition* 8 (2002): 1-19.

Nordlund, Marcus. "Consilient Literary Interpretation." *Philosophy and Literature* 26 (2002): 312-33.

---. *Shakespeare and the Nature of Love: Literature, Culture, Evolution.* Evanston: Northwestern UP, 2007.

Pinker, Steven. The *Blank Slate: The Modern Denial of Human Nature*. New York: Viking, 2002.

---. *How the Mind Works*. New York: Norton, 1997.

---. *The Language Instinct: How the Mind Creates Language*. New York: William Morrow, 1994.

---. *The Stuff of Thought: Language as a Window into Human Nature*. New York: Viking, 2007.

---. "Toward a Consilient Study of Literature." *Philosophy and Literature* 31 (2007): 162-78.

---, and Ray Jackendoff. "The Faculty of Language: What's Special about It?" *Cognition* 95 (2004): 201-36.

Plotkin, Henry. "The Power of Culture." Dunbar and Barrett, Oxford Handbook 11-19.

Richerson, Peter J., and Richard Boyd. "Darwinian Evolutionary Ethics: Between Patriotism and Sympathy." *Evolution and Ethics: Human Morality in Biological and Religious Perspective*. Ed. Philip Clayton and Jeffrey Schloss. Grand Rapids, MI: W. B. Erdmans, 2004. 55-77.

---. *Not by Genes Alone: How Culture Transformed Human Evolution*. Chicago: U of Chicago P, 2005.

Saunders, Judith. "Evolutionary Biological Issues in Edith Wharton's *The Children*." *College Literature* 32 (2005): 83-102.

---. "Male Reproductive Strategies in Sherwood Anderson's 'The Untold Lie.'" *Philosophy and Literature* 31 (2007): 311-22.

Scalise Sugiyama, Michelle. "Cultural Variation Is Part of Human Nature: Literary Universals, Context-Sensitivity, and 'Shakespeare in the Bush.'" *Human Nature* 14 (2003): 383-96.

---. Feminine Nature: An Evolutionary Analysis of Hemingway's Women Characters. Dissertation, U of California, Santa Barbara, 1997.

---. "Food, Foragers, and Folklore: The Role of Narrative in Human Subsistence." *Evolution and Human Behavior* 22 (2001): 221-240.

---. "How an Interest in Fiction Could Have Evolved." A review of *Comeuppance: Costly Signaling, Altruistic Punishment, and Other Biological Components of Fiction. Evolution and Human Behavior*, forthcoming.

---. "Narrative as Social Mapping--Case Study: The Trickster Genre and the Free Rider Problem." *Ometeca*, 12 (2008): 24-42.

---. "Narrative Theory and Function: Why Evolution Matters." *Philosophy and Literature* 25 (2001): 233-50.

---. "New Science, Old Myth: An Evolutionary Critique of the Oedipal Paradigm." *Mosaic* 34 (2001): 121-36.

---. "On the Origins of Narrative: Storyteller Bias as a Fitness-Enhancing Strategy." *Human Nature* 7 (1996): 403-425.

---. "Reverse-Engineering Narrative: Evidence of Special Design." Gottschall and Wilson 177-96.

Shapiro, Lawrence, "Do's and Don'ts for Darwinizing Psychology." *The Evolution of Mind.* Ed. Denise Cummins and Colin Allen. New York: Oxford UP, 1998. 243-259.

Sheenan, Stephen. "Evolutionary Perspectives in Archaeology: From Culture History to Cultural Evolution." Dunbar and Barrett, *Oxford Handbook* 587-97.

Shubin, Neil. *Your Inner Fish: A Journey Into the 3.5-Billion-Year History of the Human Body*. New York: Pantheon Books, 2008.

Slingerland, Edward. *What Science Offers the Humanities*. Cambridge: Cambridge UP, 2008.

Smail, Daniel Lord. *On Deep History and the Brain*. Berkeley: U of California P, 2008.

Smith, David Livingstone. *The Most Dangerous Animal: Human Nature and the Origins of War*. New York: St. Martins Press, 2007.

---. *Why We Lie: The Evolutionary Roots of Deception and the Unconscious Mind.* New York: St. Martins press, 2004.

Smith, Murray. "Darwin and the Directors: Film, Emotion, and the Face in the Age of Evolution." *TLS* 7 February 2003: 13-15.

---. "The Art of Film in the Age of Evolution," in Matthias Brütsch and Vinzenz Hediger (eds.), *Emotion and Film* (2005, University of Zurich).

Sober, Elliott, and David Sloan Wilson. *Unto Others: The Evolution and Psychology of Unselfish Behavior*. Cambridge: Harvard UP, 1998.

Spolsky, Ellen. "Frozen in Time?" Review of *The Literary Animal: Evolution and the Nature of Narrative*. Eds. Jonathan Gottschall and D.S. Wilson. *Poetics Today* 28 (Winter 2007): 807-816

---. *Gaps in Nature: Literary Interpretation and the Modular Mind*. Albany, NY: SUNY P, 1993.

Steen, Francis. "The Paradox of Narrative Thinking." *The Journal of Cultural and Evolutionary Psychology* 3 (2005): 87-105.

Sterelny, Kim. "An Alternative Evolutionary Psychology?" Gangestad and Simpson, *Evolution of Mind* 178-85.

Storey, Robert. *Mimesis and the Human Animal: On the Biogenetic Foundations of Literary Representation*. Evanston: Northwestern UP, 1996.

Stout, David B. "Aesthetics in 'Primitive Societies.'" *Aesthetic Theories: Studies in the Philosophy of Art*. Ed. Carol F. Jopling. New York: Dutton, 1971. 30-34.

Swirski, Peter. *Of Literature and Knowledge: Explorations in Narrative Thought Experiments, Evolution, and Game Theory*. London: Routledge, 2007.

Symons, Donald. *The Evolution of Human Sexuality*. Oxford: Oxford UP, 1979.

---. "On the Use and Misuse of Darwinism in the Study of Human Behavior." Barkow, Cosmides, and Tooby 137-162.

Tinbergen, Niko. "On Aims and Methods of Ethology." *Zeitschrift für Tierpsychologie* 20 (1963): 410-433.

---. *The Study of Instinct*. Oxford: Clarendon, 1951.

Tomasello, Michael. "The Human Adaptation for Culture." *Annual Review of Anthropology* 28 (1999): 509-29.

Tooby, John, and Leda Cosmides. "Does Beauty Build Adapted Minds? Toward an Evolutionary Theory of Aesthetics, Fiction, and the Arts." *SubStance* 30 (2001): 6-27.

---. "The Past Explains the Present: Emotional Adaptations and the Structure of Ancestral Environments." *Ethology and Sociobiology* 11 (1990): 375-424.

Trivers, Robert L. "Introduction." *Self-Deception: An Adaptive Mechanism?* Ed. Joan S. Lockhard and Delroy L Paulhus. Englewood Cliffs, NJ: Prentice Hall, 1988. vii-ix.

---. "Parental Investment and Sexual Selection." *Sexual Selection and the Descent of Man 1871-1971*. Ed. Bernard Campbell. Chicago: Aldine, 1972. 136-79.

---. *Social Evolution*. Menlo Park, CA: Benjamin/Cummins, 1985.

Turner, Frederick. *Natural Classicism: Essays on Literature and Science*. 1985. Charlottesville: University of Virginia Press, 1992.

Wilson, David Sloan. "Evolutionary Social Constructivism." Gottschall and Wilson 20-37.

---. *Evolution for Everyone. How Darwin's Theory Can Change the Way We Think about Our Lives*. New York: Delacorte, 2007.

---. "Evolution, Morality, and Human Potential." *Evolutionary Psychology: Alternative Approaches*. Ed. Steven J. Scher and Frederick Rauscher. Boston: Kluwer, 2002. 55-70.

---. "Group-Level Evolutionary Processes." Dunbar and Barrett, *Oxford Handbook* 49-55.

---. and Edward O. Wilson. "Rethinking the Theoretical Foundation of Sociobiology." *The Quarterly Review of Biology* 82 (2007): 327-48.

Wilson, Edward O. *Consilience: The Unity of Knowledge*. New York: Knopf, 1998.

---. *Sociobiology: The New Synthesis, Twenty-Fifth Anniversary Edition*. Cambridge: Harvard UP, 2000.

Wilson, Timothy D. *Strangers to Ourselves: Discovering the Adaptive Unconscious*. Cambridge: Harvard UP, 2002.

Zunshine, Lisa. Zunshine. "Why Jane Austen Was Different, and Why we May Need Cognitive Science to See It." *Style*, 41 (2007): 277-99.

In Spain

Cordón, Faustino. "El darwinismo en España. En el 1er centenario de la muerte de Charles Darwin (1882-1982)." *Aportación innovadora de Faustino Cordón a la teoría evolucionista*. Barcelona: Anthropos, 1982. [includes a forty-page chronological bibliography of all publications in Spanish related to Darwin's works from 1859 through 1982]

Darwin, Charles. *De l'origine des espèces, ou des lois du progrès chez les êtres organisés*. Trans. and Preface by Clémence Royer. Paris: Guillaumin and V. Masson, 1862.

---. *Origen de las especies por selección natural: o resumen de las leyes de transformación de los seres organizados (con dos prefacios de Madame Clemencia Royer)*. Madrid: Jacobo María Luengo, 1872.

---. *Origen de las especies: por medio de la selección natural ó la conservación de las razas favorecidas en la lucha por la existencia*. Trans. Enrique Godínez. Madrid: José de Rojas, 1873.

---. *El origen del hombre: la selección natural y la sexual (primera versión española)*. Barcelona: Renaixensa, 1876. [recognized as the first Spanish translation of Darwin, later attributed to Joaquím Bartrina].

---. *Origen de las especies por medio de la selección natural ó la conservación de las razas favorecidas en la lucha por la existencia*. Trans. Enrique Godínez. Madrid: Perojo, 1877 [first complete, authorized translation of Darwin in Spanish].

---. *El origen de las especies*. Trans. Antonio de Zulueta. Madrid: Alianza, 2003.

---. *La descendance de l'homme et la sélection sexuelle*. Trans. Jean-Jacques Moulinié. Preface by Carl Vogt. Paris: Reinwald, 1872.

---. *La descendencia del hombre y la selección en relación al sexo*. Trans. José del Perojo and Enrique Camps. Madrid: Administración de la Revista de Medicina y Cirugía Prácticas (Rivadeneyra), 1885. [first complete, authorized translation of the *Descent* in Spanish].

García Camarero, Ernesto, and Enrique García Camarero, Eds. *La polémica de la ciencia española*. Madrid : Alianza Editorial, 1970.

Glick, Thomas. *Darwin en España*. Barcelona: Península, 1982.

---, ed. *The Comparative Reception of Darwinism*. Austin, TX: U Texas P, 1972.

Núñez, Diego. *El darwinismo en España*. Madrid: Castalia, 1969.

Pratt, Dale J. *Signs of Science: Literature, Science, and Spanish Modernity since 1868*. West Lafayette, IN: Purdue UP, 2001.

Sánchez Ron, José Manuel. *Cincel, martillo y piedra. Historia de la ciencia en España (siglos XIX y XX)*. Madrid : Taurus, 1999.

In Latin America

Collichio, Terezinha A. F. *Miranda Azevedo e o Darwinismo no Brasil.* Belo Horizonte: Itatiaia; São Paulo: Editora da Universidade de São Paulo, 1988.

Domingues, Heloisa, et al., eds. *A Recepção do Darwinismo no Brasil.* Rio de Janeiro: Fiocruz, 2003.

Glick, Thomas F., Miguel Angel Puig-Samper, and Rosaura Ruiz, eds. *The Reception of Darwinism in the Iberian World: Spain, Spanish America, Brazil.* Boston: Kluwer, 2001.

---. *El Darwinismo en España e Iberoamerica.* Mexico: Universidad Nacional Autónoma de México, 1999.

Lopez Cid, Maria Rosa. "O aperfeiçoamento do homen por meio da seleção: Miranda Azevedo e a divulgação do darwinismo, no Brasil, na década de 1870." Diss. Casa Oswaldo Cruz/Fio Cruz, Rio de Janeiro, Brazil, 2004.

Miranda Azevedo, Agusto Cesar de. "Darwinismo: seu passado, seu presente, seu futuro". *Conferências Populares da Frequesia da Glória.* Numbers 1-10. Rio de Janeiro: J. Villenueve S.C., 1876.

Pruna, Pedro M. *Darwinismo y sociedad en Cuba: siglo XIX.* Madrid: Consejo Superior de Investigaciones Científicas, 1989.

Waizbort, Ricardo. "Cento e quarenta anos sem Charles Darwin bastam: sobre variedades, espécies e definições." *Principia.* 4.1, (2000): 141-84.

On line

www.ometeca.org. Journal treating the relations between the humanities and the sciences in the Hispanic world. Many articles on Darwinism in Hispanic literature, including Vol. 10 no. 1, a special issue on Darwin and literature.

http://www.umsl.edu/~carrolljc/index.htm. Joseph Carroll's web site for Darwin and Literature

http://www.hbes.com/. Human Behavior and Evolution Society. A complete listing of all sources.

http://www.ling.ed.ac.uk/~jim/. Language Evolution.

http://human-nature.com/science-as-culture/. Home of Human Nature Review

http://evolution.binghamton.edu/evos/. Evolutionary Studies Program at Binghamton University. Evolution and the Humanities.

http://www.epjournal.net/index.html. Evolutionary Psychology Journal

http://www.class.uh.edu/cogsci/index.html. Cognitive Science, Humanities, and the Arts homepage

http://www2.bc.edu/~richarad/lcb/home.html.Literature, Cognition, and the Brain.

Jerry Hoeg

Dr. Jerry Hoeg is a Professor of Spanish at the Pennsylvania State University, Fayette. He received his Ph.D. in Spanish from Arizona State University

Kevin S. Larsen

Dr. Kevin S. Larsen is a Professor of Spanish and Adjunct Professor of Religious Studies at the University of Wyoming, Laramie. He received his Ph.D. in Romance Languages and Literatures from Harvard University.